LITERARY JOURNALISM
in the TWENTIETH CENTURY

LITERARY JOURNALISM
in the TWENTIETH CENTURY

Edited by
Norman Sims

New York   Oxford
OXFORD UNIVERSITY PRESS
1990

Oxford University Press

Oxford New York Toronto
Delhi Bombay Calcutta Madras Karachi
Petaling Jaya Singapore Hong Kong Tokyo
Nairobi Dar es Salaam Cape Town
Melbourne Auckland

and associated companies in
Berlin Ibaden

"The Mother of Literature: Journalism and *The Grapes of Wrath*" is used by permission.
Copyright 1989 by William Howarth. An earlier version of this chapter appeared
in *New Essays on The Grapes of Wrath*, published by Cambridge University Press.

"The Borderlands of Culture: Writing by W. E. B. Du Bois, James Agee, Tillie Olsen, and
Gloria Anzaldúa" is used by permission. Copyright 1990 by Shelley Fisher Fishkin.

"The Politics of the Plain Style" by Hugh Kenner first appeared in *The New York Times
Book Review* on Sept. 15, 1985. Copyright 1985 by Hugh Kenner.

"Artists in Uniform" originally appeared in *Harper's Magazine* in March, 1953.
"Settling the Colonel's Hash" originally appeared in *Harper's Magazine* in February, 1954.
Copyright by Mary McCarthy. Reprinted by permission.

"Unsettling the Colonel's Hash: 'Fact' in Autobiography" by Darrel Mansell is reprinted
from *Modern Language Quarterly*, Vol. 37, No. 2 (June 1976). Reprinted by permission
of the publisher and the author.

Library of Congress Cataloging-in-Publication Data
Literary journalism in the twentieth century / edited by Norman Sims.
p. cm. Includes bibliographical references.
ISBN 0-19-505964-6. ISBN 0-19-505965-4 (pbk.)
1. American prose literature—20th century—History and criticism.
2. Reportage literature, American—History and criticism.
3. Journalism—United States—20th century. I. Sims, Norman.
PS366.R44L57 1990 818'.50809—dc20 90-6732

Printing 9 8 7 6 5 4 3 2 1

Printed in the United States of America
on acid-free paper

Preface

> Borders are set up to define the places that are safe and
> unsafe, to distinguish *us* from *them*. A border is a dividing
> line, a narrow strip along a steep edge. A borderland is . . .
> created by the emotional residue of an unnatural boundary. It
> is in a constant state of transition. The prohibited and forbid-
> den are its inhabitants.
>
> Gloria Anzaldúa

During the last dozen years, two of the most fertile fields of study
in nonfiction have been literary journalism and the borderlands
between fact and fiction.

The fact, obvious and verifiable, the cornerstone of journalism in
this century, has been viewed from new perspectives. First, there
was a storm of criticism in the 1960s and 1970s surrounding the New
Journalism, which challenged the assumptions of objective reporting. Then, deconstructionists in literary theory began declaring, in
their curiously technical language, that the differences between
fact and fiction writing were mere conventions. Communications
scholars in the fields known as critical studies and cultural studies—
who had long been interested in New Journalism—extended their
analysis to literary journalism before the 1960s.

Representatives from each of those areas have contributed to
this volume. We have tried to use standard English so the chapters
would be accessible to students in several disciplines including
journalism, American studies, English, and communications. We
wanted to avoid, if possible, language that demands translation.
Our broader aims are twofold: first, to establish through a few
historical studies that the New Journalism of the 1960s was preceded by a substantial American experience with literary journal-

ism in the twentieth century, and second, to bring some of the insights from contemporary theory to bear on the works of literary journalists since the early 1900s.

In the twentieth century, literary journalists developed a working tradition. They reshaped literary styles to permit passages across the borders between fact and fiction, journalism and autobiography, and reporting and sociology in such a way that their readers' expectations and confidences were not violated. For example, George Orwell wrote in what Hugh Kenner calls the "plain style," a personal, honest, casual presentation that avoided raising questions. British literary theorist Ian Watt once noted that journalism is written so the reader never asks, "Who made this up?" Orwell mastered the plain style so well that decades passed before Kenner asked the question. Fortunately, we can now talk about Orwell's style as a form of literary presentation, rather than assuming that style and reality are mirror images.

Much of the recent scholarship in journalism and communications has focused on discussions of objective fact, and on the relationship of texts to reality. In this volume, we are not as concerned with fact as we are with its presentation. Most of us share a belief that objectivity is not the heartbeat of nonfiction. If we actually verified the facts we read, there would be few problems with objectivity. Literary theorist Wolfgang Iser has suggested, correctly I think, that we do not verify what we read even when it is within our power to do so. The literary forms we encounter in texts influence our reading. We assume *The New York Times* is factual, that autobiography contains personal testimony, and that journalism—simply and honestly presented—is free of the creations and symbolism of fiction. Bringing those assumptions up for close examination has been part of the triumph of recent literary history. We are not denying the existence of fact or the importance of objectivity in some forms of journalism. We are instead exploring topics of equal importance in the analysis of literary journalism.

This volume addresses both the history of literary journalism in the twentieth century and some of the issues of literary theory surrounding it. Several important figures in the development of literary journalism are profiled, such as Hutchins Hapgood, Ernest Hemingway and Joseph Mitchell. In addition, several chapters are devoted to the styles and strategies writers developed while extending literary journalism into the borderlands of fiction, memoir, and sociology.

Part I contains historical studies from the early 1900s through the arrival of the New Journalism in the 1960s. Tom Connery writes about Stephen Crane, Hutchins Hapgood, Abraham Cahan, and Lincoln Steffens, pioneers in literary journalism early in the century, and shows that many of the theoretical issues were recognized even then. Ron Weber examines the extended narrative nonfiction written by Ernest Hemingway and the motives that led such a talented fiction writer to return to fact writing in *Death in the Afternoon, Green Hills of Africa,* and other works. Weber touches on the major forms of nonfiction: Hemingway wrote memoirs, travel articles, and two narrative books on bullfighting. William Howarth addresses a question that has troubled me for several years: Why did John Steinbeck write *Grapes of Wrath* as a novel, when he had a wealth of journalistic material? Howarth carefully explains the influence of Depression-era documentary films and photography on one of the greatest novels of the century. While fiction had a profound influence on nonfiction writers, Howarth demonstrates that the flow of influence also ran the other way. My contribution on Joseph Mitchell of *The New Yorker* deals with one of the true symbolists among literary journalists. Mitchell has avoided talking about his works for nearly twenty-five years, yet he and his close friend, A. J. Liebling, made enduring contributions to the history of literary journalism. Finally, John Pauly brings a different approach to the topic of the New Journalism, arguing it was as much "a style of cultural politics" as a literary form.

Part II contains critical discussions of the fact–fiction border. Shelley Fisher Fishkin shows that literary journalism need not be limited to works by mainstream writers. Her study uses W. E. B. Du Bois, James Agee, Tillie Olsen, and Gloria Anzaldúa as examples of those silenced by race, class, or sex, and those who searched for ways to express an existence outside of the dominant paradigm. Hugh Kenner labels the literary form of truth-telling as the plain style. He mentions writers such as Swift, Defoe, Mencken, and Orwell, who also happened to be, at times, literary journalists. In an artful essay, Kenner describes our contradictory feelings about the plain style of journalism. David Eason draws a distinction between realist and modernist writers—between Tom Wolfe and Joan

Didion, for example. The categories are valuable tools of thought because they describe philosophical differences brought to the task of writing not only by New Journalists but by all the literary journalists in the twentieth century. Most literary critics find the modernists more worthy of attention than the realists. Kathy Smith, however, shines the light of contemporary literary theory on John McPhee, one of the most esteemed of realists. With a perspective rooted in the works of Roland Barthes, Hayden White, and others, she opens McPhee's texts in exquisite sections like a Cubist painting. If the insights of literary theory apply to a realist like McPhee, then perhaps all journalism, even the newspaper, can be seen from this perspective.

Part III reprints a classic episode from the 1950s involving one of America's best known novelists and essayists, Mary McCarthy. The first article, "Artists in Uniform," published in *Harper's* in March, 1953, tells a story about a woman who encountered an anti-Semitic colonel while on a train ride. The second article, "Settling the Colonel's Hash," reports McCarthy's experiences after the first article appeared. She was unsure whether the first article was "a piece of reporting or a fragment of autobiography," but readers had taken it as a fictional story. They saw symbolic meaning in what the colonel and the narrator ate for lunch, the clothes the woman wore, and that she found herself sitting with nuns in the passenger car. McCarthy, the fiction writer, was thrown into an odd position. She found herself arguing that the story was nonfiction and should be interpreted differently than a short story: "There were no symbols in this story; there was no deeper level," she said. The last contribution in the section, by Darrel Mansell, takes issue with McCarthy's assumption that two separate literary arenas exist: "autobiography" and "narrative fiction." Northrop Frye once noted, Mansell says, that "an autobiography coming into a library would be classified as nonfiction if the librarian believed the author, and as fiction if she thought he was lying." Lacking independent means for making such judgments, we tend to rely on our assumptions about literary form, on the style of presentation. Mansell's comments on autobiography apply equally to the world of nonfiction in general. Part III serves as a case study of the twisted paths that literary journalism and autobiography have followed during the twentieth century.

Defining literary journalism presents problems. For one thing, some writers dislike the term because "literary" seems a self-congratulatory term. Yet they realize their work differs from standard journalism in approach and presentation. For another thing, such definitions are always vague. However well done, they don't necessarily help us distinguish one thing from another. In *Representing Reality,* John Warnock suggested that literary nonfiction is "a factual representation of the truth in narrative form." Ronald Weber emphasized in *The Literature of Fact* that literary nonfiction holds a value as language that reaches beyond its informative qualities. Chris Anderson in *Literary Nonfiction* said literary essays and journalism can be informative, reflective, and personal. My colleague Jim Boylan defines literary journalism as a form "that aims at substantial literary quality and fidelity to the truth as the writer sees it; it is writing that seeks to encompass aspects of life and culture that may lie beyond the grasp of other forms of journalism." In all these cases, it seems easier and more definitive to cite examples: George Orwell's "A Hanging," Joan Didion's *Salvador,* John McPhee's *Encounters with the Archdruid,* James Agee's *Let Us Now Praise Famous Men,* for example. In this book, we tend to examine authors and their works rather than attempting to settle the matter of definition. Surprising agreement exists among writers and literary theorists on the question of who creates literary journalism, much more agreement than may ever be found on a definition. I suppose the same would be true if we tried to define a "novel."

We have chosen not to reprint literary texts here, with the exception of Mary McCarthy's articles, because they are generally available in the library or the bookstore. Hemingway's nonfiction, Joseph Mitchell's books, Orwell's *Down and Out in Paris and London* and *Homage to Catalonia,* W. E. B. Du Bois's *Souls of Black Folk,* Agee's *Let Us Now Praise Famous Men,* Tillie Olsen's *Silences,* and the works of Tom Wolfe, Joan Didion, and John McPhee can all be easily located and matched to the essays in this volume. In addition, some of the contributors have written or edited valuable books on related subjects, such as Shelley Fisher Fishkin's *From Fact to Fiction* and Ron Weber's *The Reporter as Artist.* A bibliography has been included. Many of the original works on that list could be used in a classroom to illuminate the articles contained in this collection.

My thanks go out to all who helped with this book. Howard Ziff and Jim Boylan, my colleagues in the Journalism Department at the University of Massachusetts, made me aware of works by many of the contributors. William Sisler of Oxford University Press supported and encouraged our efforts. Above all, my thanks to the contributors who have in this work, and many others, sought to make sense of literary journalism.

Amherst   N.S.
January 1990

Contents

PART I

1

A Third Way to Tell the Story: American Literary Journalism at the Turn of the Century

Thomas B. Connery

In an 1896 Richard Harding Davis short story, "The Red Cross Girl," a reporter, Sam Ward, is assigned to cover the opening of a new convalescent center. The opening is news because it involves a number of rich and prominent people, including the wealthy man who donated the money for the center. Ward writes his report and turns it in, but the copy editor cannot believe what Ward has written and turns it over to the paper's managing editor, with this comment:

> "Read the opening paragraph," protested Collins. "It's like that for a column! It's all about a girl—about a Red Cross nurse. Not a word about Flagg or Lord Deptford. No speeches! It's not a news story at all. It's an editorial, and an essay, and a spring poem. I don't know what it is. And, what's worse," wailed the copy editor defiantly and to the amazement of all, "it's so darned good that you can't touch it. You've got to let it go or kill it."[1]

Elliot, the managing editor, decides to run the article as it is, over the continued objections of Collins:

> Collins, strong through many years of faithful service, backed by the traditions of the profession, snorted scornfully.

"But it's not news!"

"It's not news," said Elliot doubtfully; "but it's the kind of story that made Frank O'Malley famous. It's the kind of story that drives men out of this business into the arms of what Kipling calls 'the illegitimate sister.' "

As literary art, the story has little merit. As a cultural artifact, however, the story is rich in symbolic significance, illustrating the very nature of newspaper journalism as an information medium and literary genre in urban society as the new century was beginning. The story suggests that the concept of news, which had been evolving in its modern form since the 1830s, was already becoming narrowly defined, established, and even traditional. It also depicts a popular conflict within the newspaper, between reporter and copy editor, with the reporter's artistic perception and ability fighting against the hack copy editor who wants to reduce all newspaper articles to a predictable formula. Finally, the story alludes to a fact–fiction tension that had become particularly evident in the two dominant American prose forms as modern journalism and realistic fiction both emerged in the second half of the nineteenth century. The line was being drawn between two distinct prose categories, with the news (facts/nonimaginative writing), that world of real people and actual events, the legitimate offspring of printed prose on one side, and the novel or short story (fiction/imaginative writing), that realm of created people and events, the illegitimate sister, on the other.[2]

The realistic movement that had made the common and ordinary legitimate topics for writers had been at least partially fueled by a cultural need to know and understand the rapidly changing world, and by a staunch faith that reality was comprehensible through printed prose. Around the turn of the century in the United States a belief that reality could be identified and objectified came to literary expression in its purest form in the journalistic news story, in magazines through the crusading, fact-filled articles of the Muckrakers, and in fiction through stories that claimed to mirror life being lived. But while there would appear to be an intersecting of journalism and fiction that increased the fact–fiction tension, the desire to keep journalism objective, noninterpretive, and noncreative in form and content remained ascendant, and storytelling increasingly came to be synonymous with fiction.[3]

Obviously the boundaries of these two distinct prose forms were not merely marked by one being fact and the other fiction, but by strictures of style and substance as well. The copy editor in Davis's short story calls the reporter's article "an editorial, and an essay, and a spring poem," before finally conceding, "I don't know what it is." That is, the article did not seem to be fiction because the people in it were real and the events had just happened. It did not seem to be journalism because it did not contain the type of facts that made conventional news and because in its presentation it tended to interpret events through its narrative point of view.

If the common view of the journalistic establishment was typified by the attitude of the copy editor in "The Red Cross Girl," the literary establishment's stance was best expressed by H. W. Boynton, a writer and literary critic who declared that "journalism has, strictly, no literary aspect. The real business of journalism is to record or to comment, not to create or interpret."[4] Yet "The Red Cross Girl" also illustrates a resistance to restricting the form and content of newspaper journalism, and suggests that two categories of printed prose to depict observed life were not enough, but a third—a literary journalism—was possible and necessary.

At least since print technology made relatively rapid and widespread dissemination of information and narrations possible, writers have grappled with finding the most effective form of prose discourse to make life comprehensible.[5] Because of this quest for form, much of post-Gutenberg literary history, including stylistic development in journalism, has been marked by attempts to find the appropriate way to express what has been witnessed. While the direction this search has taken has been a reflection of particular cultures and eras, the search has been continual. It usually has signified a persistence in discovering ways to relate language and text, and a desire to discover the limitations of the printed word in recording and depicting reality.

Davis's reporter was representative in the sense that despite the enormous cultural pressure to classify and separate prose into creative and noncreative categories, resistance to such sharp demarcations was salient at the turn of the century. That resistance and the writing that resulted can be identified, characterized and analyzed as part of this on-going news–novel discourse.

The concern here, however, will not be with the news–novel

discourse but with the nature of an element of that discourse—literary journalism. This is not an attempt at genre making, nor is it an attempt to anoint a type of journalism as literary art. Rather the purpose is to provide a historical context, as well as a journalistic–literary perspective for a form of printed prose that has been a significant form of cultural expression in the twentieth century but has been either ignored, mislabeled, or misread.

As already noted, a literary journalistic account did not just record and report, it interpreted as well. It did so by subjectively placing details and impressions no longer considered appropriate for the standard newspaper article into a storytelling form that was also being cast aside by the institutionalized press. In this way, it gave readers another version of reality, an interpretation of culture different from that of either most conventional journalism or most fiction that contained elements of both.

Like its descendant of the 1970s and 1980s, literary journalism in the late nineteenth and early twentieth centuries often contained factual information common in conventional journalism, but focused on presenting impressions, details, and description not central to the typical conventional newspaper report. In the process this created a different context as well. That is, such writing did not simply present facts, but the "feel" of the facts, or, as one critic has said of Stephen Crane's newspaper pieces, "a rendering of felt detail."[6]

Although some literary journalism bore a striking resemblance to what was coming to be known as newspaper feature writing, it eschewed the evolving formula of that article type, with its predictability and clichés, and naturally appropriated the narrative conventions of realistic fiction and storytelling to portray daily life as observed by the writers. In doing so, literary journalism often reconciled fact and fiction, reality and language, by being a mode of expression more imaginative than conventional journalism but less imaginative than fiction. Literary journalism, however, was less fully realized than much fiction, and was more episodic as well.

These characteristics become evident when one looks at the comments and work of the writers of literary journalism. While an inclusive investigation of the form is beyond the scope of this article, it would be worthwhile to briefly consider three different attempts at producing this third way of ordering reality: Crane's articles on New York City, Lincoln Steffens' experiment with literary journalism at the *Commercial Advertiser* newspaper, and Hutchins Hapgood's articulation of a concept of literary journalism that called for nonfiction novels.[7]

STEPHEN CRANE

Crane's nonfiction renderings of New York City life consist of articles commonly called "The New York City Sketches." In many of these articles, most of which were published in the *New York Press* in 1894, Crane took relatively insignificant incidents—for example, a wagon breaking down on a busy street—and transformed them into interpretations of city life that revealed aspects of human nature.[8] To depict the reality he had witnessed, and specifically the behavior of New York's people, Crane used a host of literary techniques, including contrast, dialogue, concrete description, detailed scene setting, careful word selection that built a repetition of imagery, and irony.

For instance, in "When Man Falls, A Crowd Gathers" it is the end of the work day and an Italian man and his companion, a young boy, are among the crowd of New Yorkers hurrying home when the man falls to the pavement in an epileptic fit and a crowd immediately forms around the two. A man collapsing on a city street might be a small news item in some newspapers. Crane was not concerned with the facts of an incident, but with how people were affected by the incident and how they behaved in particular situations.

Crane's account shows New Yorkers who complicate a potentially tragic situation rather than help resolve or improve it. At the same time, he indirectly suggests that the kind of factual information associated with conventional news reporting is irrelevant to the point of being "abstract":

> Meanwhile others with magnificent passions for abstract statistical information were questioning the boy. "What's his name?" "Where does he live?"[9]

That is not the kind of information Crane intends to provide his readers. The lead paragraph begins in a traditional storytelling manner, presenting a close-up view of the evening urban land-scape, and a particular scene in that landscape by immediately focusing on the old man and the boy:

> A man and a boy were trudging slowly along an East-Side street. It was nearly six o'clock in the evening and this street, which led to one of the East River ferries, was crowded with laborers, shop men and shop women, hurrying to their dinners. The store windows were a-glare.
>
> The man and the boy conversed in Italian, mumbling the soft syllables and making little quick egotistical gestures. They walked. . . .[10]

Crane then creates a scene that borders on chaos, and establishes the crowd as a threatening body that jostles the boy but provides no assistance. Even a policeman, the symbol of control and authority in the city, arrives late and proves clumsy and ineffective. When an ambulance finally arrives, Crane displays two characteristics of his writing: color imagery to depict a scene and ironic contrasting. First, there is the sound of the ambulance "out of the golden haze made by the lamps far up the street," and then

> the black ambulance with its red light, its galloping horse, its dull gleam of lettering and bright shine of gong clattered into view. A young man, as imperturbable as always as if he were going to a picnic, sat thoughtfully upon the rear seat.[11]

Thus, readers were not simply informed in a paragraph or two that a man had collapsed on the street. Instead, Crane carefully selected and organized his facts and observations to present his subjective view of the incident as he created a human, urban context. Crane once said that he wrote "to give readers a slice out of life," and that when he wrote *Maggie* he had no other purpose "than to show people as they seem to me." He did both with his literary journalism. Nevertheless, that newspapers were not quite sure how to classify and present such accounts is evident in the article's subhead: "A Graphic Study of New York."[12]

Such "studies" by Crane and others may be what some critics of that period referred to when they talked about a "higher journalism" or a "transfigured reporter." For instance, Boynton, who declared journalism was never creative or interpretative, admitted that such "pure" journalism was not common and that at times a writer's genius emerged and interpretation occurred. With the best such efforts, Boynton maintained, a writer ceases "to be a machine or a mouthpiece, and becomes a 'creative' writer," with the result being a "higher journalism," or a literary journalism.[13]

Another critic, Gerald Stanley Lee, also acknowledged the literary–journalistic connection, recognized a form of writing between journalism and fiction, and tried to use the literary terms of his time to describe this writing. Lee argued in 1900 that the problem with daily journalism was not that it dealt with "passing things," but that it dealt with them in a "passing way." Lee said the average reporter asked, "What do people want?" because the reporter survived that way, and therefore was not interested "in the passing for what it is."[14] Lee also claimed that there was such a journalist as a "poet reporter" who could "report a day forever, to make a day last so that no procession of flaming sunsets shall put that day out." According to Lee, the daily journalist could become a "transfigured reporter," or a reporter "who is more of an artist than artists, an artist who is more of a journalist than the journalists."[15]

Even as they recognized a third form of writing, critics such as Lee and Boynton obviously were struggling with how to respond and classify such writing because it did not fit within the emerging categories. Because such writing seemed to be more literary, they ascribed rather inappropriate terms to it. What is significant is that Lee's "transfigured reporter" interpreted daily events, as the writer of Boynton's "higher journalism" might have done, as Davis's "Red Cross" reporter did, and as Crane did in his "studies." They, and literary journalism, gave "passing things" meaning and context by going beyond the facts, beyond the expectations and requirements of daily journalism and newspaper copy editors.

Unlike Lee and Boynton, Hutchins Hapgood and Lincoln Steffens, who hired Hapgood at the *Commercial Advertiser,* perceived the possibilities of such a form of writing, envisioned a philosophy or theory of literary journalism, and attempted to enact their philosophy.

LINCOLN STEFFENS

Steffens wanted writing like Crane's "graphic studies" throughout his newspaper and without an editor's qualification. He tried to impose his philosophy of reporting on the day-to-day running of a newspaper at the *Advertiser,* where he tried "to make a new kind of daily journalism, personal, literary and immediate."[16] According to Steffens, this meant covering the city in such a way that the readers could "see" people and events. He said he wanted New York reported "so that New Yorkers might see, not merely read of it, as it was: rich and poor, wicked and good, ugly and beautiful, growing, great."[17] Even if a crime occurred, Steffens said he would not rush one of his reporters to the scene just to get the news first. Rather, he claimed, he would instruct the reporter this way:

> Here, Cahan, is a report that a man has murdered his wife, a rather bloody, hacked-up crime. We don't care about that. But there's a story in it. That man loved that woman well enough once to marry her, and now he has hated her enough to cut her all to pieces. If you can find out just what happened between that wedding and this murder, you will have a novel for yourself and a short story for me. Go on now, take your time, and get this tragedy, as a tragedy.[18]

Steffens emphasized that his reporters were not expected necessarily to beat the other papers in getting the news, but in presenting the news. Thus, Steffens was not just hoping that occasionally the "transfigured reporter" might appear; rather, he was demanding it from his staff. He wanted the "transfigured reporter," the literary journalist, to be the rule rather than the rare exception.

Steffens tried to put his ideas into play at the *Commercial Advertiser* by hiring college graduates who had demonstrated some literary potential. "They were picked men and women, picked for their unusual literary prose," he said, adding that he had let certain schools—Harvard, Yale, Princeton, and Columbia—know his paper wanted "not newspaper men, but writers."[19] Among those attracted to the *Commercial Advertiser* were Carl Hovey, Steffens' star reporter, who later became editor of *Metropolitan Magazine;* Abraham Cahan, a Russian Jew who had published a novel, *Yekl,* and who became editor of the *Jewish Daily Forward;* Norman Hapgood, a respected journalist and theater critic who would later edit a Muckraker magazine; Neith Boyce, a novelist and short story writer; Eugene Walter, a playwright; Robert S. Dunn, a novelist whose former philosophy teacher, George Santayana, would visit him in the newsroom; Josiah Flint, the travel writer and recorder of hobo life; and Norman Hapgood's brother, Hutchins, whose accounts of immigrant life for the paper were published in 1902 as *The Spirit of the Ghetto,* and reprinted in 1967.[20] Hapgood later wrote that he joined the staff of the *Commercial Advertiser* because he was convinced that the paper under Steffens was conducive to his own talents and temperament, and that he would fit in well with other staffers who believed in "Steffens's idea of a literary journalism."[21]

A *Commercial Advertiser*-style of writing did not develop, but formula writing was disdained, and articles containing dialogue, or extended conversation, and scene setting, were characteristic. For example, it was common in the 1890s to let a speaker tell his or her own story, so that many articles would consist of paragraphs of quotations, allowing the narrative movement of the piece to be provided by the speaker. Although that made the story another step removed from the reader, it was effective, perhaps even more real than if the speaker had been concealed in a third-person narration because the writer thus conveyed to the reader that he or she actually had talked to the speaker. This technique gave the articles a stronger basis in fact, yet, if used properly, could create the "literary" approach Steffens called for. Consequently, the writers at the *Commercial Advertiser* frequently used this technique in all types of articles.

For instance, a piece by Adelaide Lund, with the headline, "Told in the Guard Room," begins this way:

> "For life,"
> He answered without moving his lips. He was waiting for tools to repair an electric fixture and stood a gray and black statue, facing the wall. Little muscles showed above the sagging shirt-band; loose, hopeless hands, a prison face, a voice of dead notes.
> "Will you tell me?"
> The gray image did not move.
> "I have permission to talk with the inmates."
> He lifted his eyes.
> "It's for murder." He wet his lips. "That is, they *called* it that; and I've been here nine years. . . ."[22]

The body of the article consists of the prisoner telling his own story about how he and his wife had struggled to survive at a knitting mill, how his baby son died because it was not strong enough to endure being taken to the mill daily so the mother could work, and how he murdered a lawyer who swindled him out of a patent. Here is the conclusion:

> "That's all; I'm here for life and Mary is still living and is in the poorhouse. But they are good to her. She always thinks I am coming back and has had her bed pulled up to the window, and she has sat here for nine years watching the road, and she'll die sitting there. Of course, sometimes I dream—"
>
> The guard motioned him to begin work, and in an instant he faced the wall.[23]

The story is no longer than countless others that appeared in New York's newspapers, yet because of the carefully chosen words that introduce and end it, it has a strength that most of the others do not have. Lund does not simply give her readers the prisoner and let him talk. Rather, she gave them a man who had become virtually lifeless; he had become a part of the prison wall he was repairing. Only slowly did he come to life to tell his tale, only to become one with the wall again at the end. The article is somewhat melodramatic, somewhat sentimental, and scarcely a solid vignette by fiction's standards, but the manipulation of facts to create an image was not typical of mainstream journalism and the result was much more than covering a "passing event" in a "passing way."

"All For A Flower" is another *Commercial Advertiser* example of literary journalism, and was probably written by Hutchins Hapgood. The article follows an old man as he makes his way up Broadway to a flower shop where he presents the young female clerk with newspaper clippings about the care and planting of flowers, and in return receives a carnation for his button hole. The piece contains little dialogue, relying instead on description and the man's movement to give the brief sketch direction and unity. From the beginning, the focus is entirely upon this shabby old man:

> Not many on Broadway noticed him. He was used to that, and his face had lost the look of one who may meet friends on a thoroughfare. He was an old man, slight of figure and thin, and his face was blue and his hands shrunken. But his short trousers and coat, bare

at the seams, were carefully brushed, and so was his too-large hat. He carried a heavy stick which tapped the pavement as a little old man's will, and did not thump upon it as does a larger man's. His thin gray hair, carefully combed, fell on his shabby collar.[24]

The man's poverty is further emphasized in the flower shop where flowers are being purchased for "more money . . . than he had the whole year." Receiving the flower from the clerk means so much to the man that he spends his free time collecting old newspapers and going through them in search of items on flowers and gardens.

This sketch of one lonely, old man who is made less lonely by a slight human contact and by the gift of a single flower, reads like a fictional sketch, yet readers are aware of the presence of a reporter, or observer, a "somebody who followed him, curious to see what this shabby fellow would be about in such a place." When the old man leaves the florist, the observer questions the clerk, who does not know the man's name or address, but who does explain what started the man's visits. Because the article appeared in a newspaper, and because of the presence of the reporter questioning, observing, trying to get more information, readers knew they were not reading something created by the imagination, but about a real person who lived in their "town," New York City.[25] The article depicted a very tiny part of the human condition, nothing more and nothing less, and as such it succeeded. It was not the "news" demanded by most newspapers, nor was it the more elaborate fictional short story required by magazines. It was as though readers had been given a window on New York and after having swept their eyes up and down Broadway, they trained them on the old man and followed him as he went through the motions of living. This was one way in which Steffens enabled his readers to "see" and not just hear about the city.

The same issue of the paper contained an excellent example of what Steffens called "descriptive narrative."[26] The article concerned a murder in the Italian section of the city. The facts of the case are woven into a story and consequently become secondary to the tale of the people involved. The article did not begin with the fact that a murder had taken place, but instead attempted to pull the reader into the story with narrative action before giving hard facts:

Sergeant Gray and the wardman, Farrell, were gossiping across the desk in the Leonard street police station last night when they heard two pistol shots in an alley next door. Farrell leaped for the door and Gray touched the gong which called out the reserves. Farrell's coming sent three men flying down the unlighted street, and he gave chase with some of the reserves after him. As they ran, the three, the detective and the patrolmen, strung out in a line, one of the fugitives turned and fired at Farrell. He sprinted and caught the shooter. The patrolmen passing on captured the two other men.

They were Italians. Brought back to the station they were stood up over a fourth Italian, who lay bleeding, from two wounds in the head and breast, on the floor.

"Why did you do that?" asked Sergeant Gray.

One of the three stepped forward, pointed at the wounded man who looked up at him, and said:

"He robbed our mother."

Everybody waited, watching the wounded man. He did not deny the charge. The three were locked up. Their victim was taken to the Hudson Street Hospital, where the surgeons said today that he could not live. It is a story of Little Sicily.[27]

Although such a narrative approach does not seem striking, and was not uncommon before the 1880s and 1890s, what follows is an example of what had already become the preferred news narrative form:

Walter Johnson of Gloucester, Mass., this noon shot and killed Miss Carrie Andrews in the Warren building on Park street. He then shot himself in the head after which he slashed his wrist with a razor, dying in a few minutes. The shooting was the result of a love affair. Johnson and the Andrews girl had been engaged up to within a month but for some reason, at present unknown, the engagement was broken and Johnson had been brooding over the matter since.

Miss Andrews was a pretty girl, 18 years old, and lived in Essex. The shooting occurred in the parlor. . . .[28]

The difference between the two approaches is obvious, but the significance of the difference is worth noting. In the second example, specific kinds of facts dominate: modern journalism's classical formula of who, what, where, when, and how, with a superficial nod to why, dictates the form. In the *Commercial Advertiser* article, however, storytelling and shaping the facts into a dramatic slice of New York City life are important as the writer presents not the facts of a murder, but another sad incident in the recurring saga of immigrants struggling to make their lives better. The story tells of how the Domenico Amoroso family came to the United States, worked hard, saved its money, and accumulated a $200 dowry for the only daughter, who was engaged to Felippe. The money, however, was stolen by a boarder, Salvatore, who denied the theft and left the house. Felippe agreed to marry the daughter without the dowry. All seemed well until Felippe and the girl's brothers, Giuseppe and Antonio, began to talk more and more of the stolen money. They saw Salvatore, the former boarder,

> coming and going in good clothes, with a watch and a heavy chain, rings, and beautiful pins. He worked no more. He was a gentleman of leisure. So the three panted for revenge.
>
> Last night they set out to find the fellow. They came upon him in a wine cellar, drank with him and went out on a carouse together. In the course of their walk they enticed him into the dark alley next to the police station and by and by there were the two shots, the chase and the scene in the station house.
>
> It was Salvatore who lay on the floor. Giuseppe who accused him, Antonio and Felippe who stood by and were arrested. (End of article.)

The piece went beyond the "news" but remained within nonfiction parameters and tried to show the meaning and significance of the murder, and became, as the writer put it, "a story of Little Sicily."

HUTCHINS HAPGOOD

Hutchins Hapgood described his own *Commercial Advertiser* writing not as storytelling but as "something like the *feuilleton*—a short article which is a mixture of news and personal reaction put together in a loose literary form."[29] In a 1905 *Bookman* article, Hapgood carried this a considerable step further when he specifically called for a form of writing that would combine aspects of journalism and literature.

In that article, Hapgood complained that American literature was lacking force and vitality, and that the imagination was not keeping pace with the reality of an industrialized, urbanized, and imperial America. Later in his life Hapgood described what he thought was wrong with the contemporary novel:

> Characters in novels, no matter how vivid and real they may be made to appear, have always seemed to me to lack the final touch of realized illusion. I have always proceeded on the theory that things exist in nature, which, if found, carry conviction far more complete than any so-called work of the imagination. The modern photography proceeds on this hypothesis. The art of the Photographer is to find this thing already existing, and then Photograph it—some intense form worked out in the activities of complicated natural elements.[30]

Hapgood argued in the *Bookman* article that literature could be transfused with new life if it borrowed an essential tool of the daily newspaper reporter: the interview. If the interview were employed properly, maintained Hapgood, then contemporary writing could be "carried into literature" by developing into "autobiography of an unconventional kind."[31] If the writer would only use his imagination in reconstructing a real personality rather than "imagining a character," then Hapgood said, "A section of life would be thus portrayed and a human story told."[32]

Hapgood thus envisioned the same form of writing more recently espoused by the New Journalists, a writing that would read like fiction but be about "real" people and events. He further foreshadowed recent New Journalists or literary journalists by insisting that a simple interview would be inadequate and writers would instead have to "become identified for the time being with their (the subjects of the writing) lives."[33] Today this is called *saturation* or *immersion reporting* and is a central technique of many literary journalists.

Hapgood had already practiced this to a degree when, with the help of his friend and colleague, Abraham Cahan, he immersed himself in the Jewish ghetto to produce his series of articles for the *Commercial Advertiser.* Most newspaper articles on immigrant neighborhoods appeared when something horrible or extraordinary happened there or as local color curiosities. Hapgood "brushed past the trivialities of journalistic 'local color,' " as Irving Howe has put it, to depict a subculture and specific community absent from conventional journalistic accounts.[34] As Moses Rischin has written, Hapgood's subjects "eluded both the muckraking journalists and the realistic novelists, for they were factually too soft for the former and too exotic or unsocial for the latter."[35]

Later, Hapgood would put his theory into effect in two book-length projects, *The Spirit of Labor* and *An Anarchist Woman,* both of which involved Hapgood nearly moving in with his subjects for an extended period, "observing them in revealing situations, noting their reactions and the reactions of others to them," as Gay Talese would describe the same reporting process in his work about seventy-five years later.[36] Hapgood thus emerges as a representative figure in this search for a third way to depict observed life. He not only provided a thoughtful articulation of ideas suggested by others of his time, but he attempted to put those ideas into practice. In doing so, he became a part of the ongoing news–novel discourse as well as a foreshadowing of the evolving literary form emerging from that discourse.

Hapgood's call for a new form of literature, however, stirred little discussion or interest, as the highly fashionable Muckrakers of the period continued to capture the interest of newspapers and magazines. The *Commercial Advertiser* approach to covering the city was neither emulated nor encouraged by the larger New York City newspapers, although several of them—the *Sun,* the *Herald,* and the yellow *World* and the *Journal*—were not quite locked into the emerging inverted pyramid style and conceivably could have moved in the direction of literary journalism. Furthermore, while the more imaginative approach to writing about actual people and events found in the work of Crane, Davis, and other writers was admired and often sought by newspapers or magazines, such writing was not considered acceptable mainstream practice for aspiring journalists.

Consequently, the evolution of journalism and fiction as distinct and separate prose forms continued, as the attitude of the copy editor in Davis's "The Red Cross Nurse" became newspaper dogma, and as most magazines came to prefer, for the most part, a content mix of fiction, political–social reporting, and essays.

On the other hand, the assignment of distinct purposes and roles to journalism and fiction as cultural forms of knowing does not deny the emergence of another type of prose that resisted such absolute separation, as has been demonstrated here.[37] The essence of this writing at the turn of the century can be captured by stating its primary characteristic: Such writing attempted to go beyond journalism's facts but stopped short of fiction's creations and sought a fusion of the role of observer and maker into a literary journalism that presented a third way to depict reality.

While evidence may not suggest a direct influence on future generations of literary journalists, it is clear that precursors existed and that this nonfiction prose form has a history and therefore a significance that has not yet been properly assessed. Such writing should not be analyzed or judged according to journalism's institutionalized standards nor by fiction's standards of creation, but by what it attempts and accomplishes as a distinct prose form that realistically depicts and interprets American culture.

NOTES

1. Richard Harding Davis, "The Red Cross Girl," *The Red Cross Girl* (New York: Charles Scribner's Sons, 1916), p. 26.

2. See Lennard Davis, *Factual Fictions: The Origins of the English Novel* (New York: Columbia University Press, 1983). Davis makes a strong case for a news–novel discourse that in the seventeenth and eighteenth centuries was characterized by attempts to reconcile fact and fiction into one form. The contention in this chapter is that this fact–fiction tension is part of an on-going news–novel discourse and is particularly pronounced in the second half of the nineteenth century because of developments in fiction and journalism, including new demands placed upon literature by the familiar twin forces of urbanization and industrialization, as well as the institutionalization of the press and the professionalization of journalists.

3. The literature on the rise of news and objectivity is extensive. Two commonly cited texts are Michael Schudson, *Discovering the News* (New York: Basic Books, 1978) and Dan Schiller, *Objectivity and the News: The Public and the Rise of Commercial Journalism* (Philadelphia: University of Pennsylvania Press, 1981). In another study, Harlan S. Stensaas analyzed articles in newspapers in New York, Atlanta, Chicago, New Orleans and San Francisco, and concluded that more than half of what was appearing in American newspapers by 1894 could be called objective and by 1914 two-thirds of the reports were objective. See Stensaas, "The Rise of Objectivity in U.S. Daily Newspapers, 1865–1934," paper presented at the American Journalism Historians Convention, St. Louis, Missouri, October 1986.

For an astute discussion of news as a form of knowledge and news as discourse, see John J. Pauly, "Reflections on Writing a History of News as a Form of Mass Culture," *Working Paper No. 6,* Center for Twentieth Century Studies, University of Wisconsin-Milwaukee (Fall 1985).

4. H. W. Boynton, "The Literary Aspect of Journalism," *Atlantic* 93: 846 (June 1904).

5. I briefly presented this notion in the same context in "Hutchins Hapgood and the Search for a 'New Form of Literature,' " *Journalism History* 13(1): 2–9 (Spring 1986).

6. Alan Trachtenberg, "Experiments in Another Country: Stephen Crane's City Sketches," *Southern Review* 10: 278 (1974). In an important and perceptive essay on the New Journalism of the 1960s and 1970s, David Eason characterizes new journalism as being either realist or modernist. The literary journalism discussed here would be of the former type, realist, which is governed by a belief that a writer can show a reader "real" life, that a writer can depict reality and make it comprehensible to the reader. See Eason, "The New Journalism and the Image-World," in this volume.

7. One can find isolated examples of literary journalism in the newspapers and magazines of the period, but some of the more prominent writers who produced more than the occasional piece of writing that can be classified as literary journalism include Julian Ralph, Theodore Dreiser, George Ade, William Hard, and Jacob Riis. In addition, much of Richard Harding Davis's early journalism at the *Philadelphia Press* and the *New York Evening Sun* can be called literary journalism.

8. See Stephen Crane, *Tales, Sketches, and Reports,* vol. VII of *The Works of Stephen Crane,* Fredson Bowers, ed. (Charlottesville: University Press of Virginia, 1973), pp. 275–373, and particularly, "The Broken-Down Van," "An Experiment in Misery," and "The Men in the Storm."

9. Crane, "When Man Falls, A Crowd Gathers," p. 347.

10. ibid., p. 345.

11. ibid., pp. 348–49.

12. Qualifying an article by labeling it was one way for the copy desk to deal with writing that did not seem to fit. But newspapers also could place such work on a page that separated it from more conventional reports. For instance, George Ade's accounts of daily life in Chicago appeared in the *Chicago Record* under the heading "Stories of the Streets and of the Town" on the editorial page.

13. Boynton, p. 848.

14. Gerald Stanley Lee, "Journalism as a Basis for Literature," *Atlantic* (Feb. 1900), p. 234.

15. ibid., p. 233. Other observers noted literary aspects of newspaper journalism. For instance, Hamlin Garland wrote that newspapers contained "sketches of life so vivid one wonders why writers so true and imaginative are not recognized and encouraged." Garland, *Crumbling Idols* (Cambridge, Mass.: Harvard University Press, 1960), p. 14. A French observer echoed Garland's view and declared that the American newspaper was "a huge collection of short stories." "Newspapers and Fiction," unsigned, *Scribner's Magazine* 40: 122 (July 1906). When Carrie Nation visited New York in 1901, *The Bookman* pointed out that the coverage in the *Evening Sun* contained "more humour and insight and clever description than in half a dozen of the typical novels which appear nowadays." *The Bookman* 14: 110–11 (Oct. 1901).

16. Justin Kaplan, *Lincoln Steffens* (New York: Simon and Schuster, 1974), p. 82.

17. Lincoln Steffens, *The Autobiography of Lincoln Steffens* (New York: Harcourt, Brace and Co., 1931), p. 317.

18. ibid., p. 317.

19. ibid., p. 312.

20. Hutchins Hapgood, *The Spirit of the Ghetto: Studies of the Jewish Quarter in New York* (Cambridge: Harvard University Press, 1967). Cahan's literary journalism has been collected in *Grandma Never Lived in America: The New Journalism of Abraham Cahan*, Moses Rischin, ed. (Bloomington: Indiana University Press, 1985).

21. Hutchins Hapgood, *A Victorian in the Modern World* (New York: Harcourt, Brace and Co., 1939), p. 140.

22. Adelaide Lund, "Stories of the Town, Told in the Guard Room," *New York Commercial Advertiser* (Oct. 1, 1898), p. 9.

23. The story could be called "impressionistic" because of the image created by the selective use of the wall and the prisoner.

24. All For A Flower," *New York Commercial Advertiser* (Feb. 10, 1900), p. 9.

25. In fiction at this time, the intrusive narrator was being eliminated to further enhance the actuality of the story, to further provide the illusion of reality. But the opposite can happen in journalism; the presence of a reporter confirms that the events covered were witnessed, and therefore were real or not simply the product of the writer's imagination.

26. Steffens, p. 242. Steffens explained that details used in descriptive narrative were often irrelevant to the typical newspaper account. He said with descriptive narrative what was printed might not be news but "only life."

27. "A Little Sicily Murder," *New York Commercial Advertiser* (Feb. 10, 1900), p. 3.

28. "Jealous Event To Murder," *New York Sun* (Feb. 4, 1894), p. 2.

29. Hapgood, *Victorian*, p. 138.

30. ibid., p. 194.

31. Hapgood, "A New Form of Literature," *Bookman* 21: 424 (1905).

32. ibid., p. 423.

33. ibid., p. 425.

34. Irving Howe, "The Subculture of Yiddishkeit," *New York Times Book Review* (March 19, 1967), p. 7.

35. Moses Rischin, Introduction, Hapgood, *Spirit of the Ghetto*, p. xx.

36. Gay Talese, *Fame and Obscurity* (New York: Dell Publishing, 1981), p. 9.

37. Literary journalism did not, of course, disappear. Literary journalism, or at least dramatic-descriptive narrative and more traditional storytelling, found expression on one level in the sensational tabloids of the 1920s, in the columns of Damon Runyon, the crime reporting in newspapers such as the New York *World-Telegram*, or publications such as *True Story*. On another level it was found first in such publications as *Time, Fortune,* and *The New Yorker,* and then in *Esquire, Rolling Stone,* and *New York* before appearing again in newspapers in the past twenty years.

2

Hemingway's Permanent Records

Ronald Weber

1

"Just why did Ernest Hemingway write a book on bull-fighting?" Malcolm Cowley began a thoughtful review of *Death in the Afternoon* following its publication in 1932.[1] Critics and Hemingway admirers have been similarly puzzled ever since by the exuberant, idiosyncratic nonfiction treatise on tauromachy. For one thing, he had already written memorably about bullfighting in the novel *The Sun Also Rises* and in the lengthy story "The Undefeated." Then there was the surprising matter of timing.

He was at the peak of his power and reputation as a fiction writer with *In Our Time, The Sun Also Rises,* and *A Farewell to Arms* published over a richly productive span of five years. The novel of love and war in Italy had also been a considerable commercial success, his first, with sales reaching 80,000 just a few months after its appearance in 1929. It hardly seemed an appropriate moment to set fiction aside in favor of a demanding work of nonfiction directed to a necessarily limited audience. Yet set it aside he did, laboring for two years over the work he would describe with casual understatement as a "rather technical book."[2]

Hemingway had begun his career as a newspaper journalist, so *Death in the Afternoon* could be seen as a return to his beginnings as a fact writer. Critics have usually examined his newspaper days in light of influences on the style and subject matter of the later fiction; a more obvious influence is on his work in extended nonfiction that he began with the bullfight book. Hemingway's journalistic background, however, goes only so far in accounting for the interest in fact writing that led not only to *Death in the Afternoon,* but *Green Hills of Africa, The Dangerous Summer, A Moveable Feast,* and a still unpublished account of a second African safari.

On the second page of the bullfight book he drew a distinction between the journalism he had written for newspapers in Kansas City and Toronto and what he was trying to accomplish in both his fiction and his serious nonfiction. "In writing for a newspaper," he said, "you told what happened and, with one trick and another, you communicated the emotion aided by the element of timeliness which gives a certain emotion to any account of something that has happened on that day." His goal in his writing, as opposed to his journalism, was to created work that "would be as valid in a year or in ten years or, with luck and if you stated it purely enough, always."[3] Over and over again in his remarks about journalism he returned to the same distinction: journalism was fleeting, while writing was permanent. He was willing to grant that journalism was "not whoring when done honestly with exact reporting," yet there was no escaping the hard truth that "if it was reporting they would not remember it."[4]

His aim in his extended nonfiction was always to create books that would endure, and in this sense the work was conceived in sharp opposition with his journalistic background. When he began *Death in the Afternoon* he informed his editor at Scribner's, Max Perkins, that the book would possess "a certain permanent value."[5] Some thirty years later he told the bullfighter Antonio Ordóñez that with *The Dangerous Summer* he wanted to create a "permanent record" that would endure after they both were gone.[6] Just as he sought to make lasting fictions, in his extended nonfiction he sought lasting records—and through the same means as the fiction.

As Hemingway saw it, fiction had its source in the recollected and observed facts of experience, but if the work was to last the material had to be intensified through invention into a new and independent reality. "Writing about anything actual was bad," he had Nick Adams declare in a discarded section of "Big Two-Hearted River." "It always killed it. The only writing that was any

good was what you made up, what you imagined. That made everything come true."[7] The same aesthetic principle applied to nonfiction: to escape the death of topicality fact had to be mingled with
invention, what was true with what was made up. The territory of
Hemingway's serious nonfiction was always the muddled area between fact and fiction.

Although it is a mistake to think of his nonfiction books as
simple records of fact, it is equally mistaken to consider them
fiction by another name. When George Plimpton, questioning him
for a *Paris Review* interview, lumped together as novels *Green
Hills of Africa, To Have and Have Not,* and *Across the River and
Into the Trees,* Hemingway was quick to straighten him out:

> No, that is not true. The *Green Hills of Africa* is not a novel but was
> written in an attempt to write an absolutely true book to see
> whether the shape of a country and the pattern of a month's action
> could, if truly presented, compete with a work of the imagination.

He added that two stories that had come from his African material,
"The Snows of Kilimanjaro" and "The Short Happy Life of Francis
Macomber," had been "invented from the knowledge and experience acquired on the same long hunting trip," but that the book
about one month of the adventure had been a "truthful account."[8]
The nonfiction books were each approached in similar fashion as
truthful accounts of experience, meaning that each was composed
of more fact than fiction.

Hemingway may have turned his hand to nonfiction with *Death
in the Afternoon* out of a wish to enter the ranks of the men of
letters, those wide-ranging professionals who moved easily from
stories and novels to travel books and literary criticism, essays and
memoirs. During the 1930s he would follow the bullfight book with
a variety of work—magazine articles, a play and a film, stories and
novels, and another ambitious work of nonfiction. He even mentioned to Max Perkins during the period that he thought he would
put out a collection of his essays as well as his stories.

Whatever the larger motivation, it is clear that he had long
wanted to do a major instructional book about bullfighting. In his
first letter to Perkins on April 15, 1925, while he was still under
contract to another publishing house, he mentioned as a future
project "a sort of Doughty's Arabia Deserta of the Bull Ring, a

very big book with some wonderful pictures."[9] When the editor heard no more about the project for a year and a half, he gently prodded his rising star, now in the Scribner fold, about what "sounded like a most interesting and individual work." He quickly added that he was certain that Hemingway "meant to make its preparation a matter of years."[10]

In fact, Hemingway had been gathering material for the book since his arrival in Europe as a foreign correspondent. In 1923 the *Toronto Star Weekly* carried a long lead article called "Bull-Fighting is Not a Sport—It Is a Tragedy" drawn from a trip he had taken to Spain that spring. A week later the newspaper carried "World Series of Bull Fighting a Mad, Whirling Carnival," an account of the best bullfight he had seen during Pamplona's annual festival of *San Fermín*. Over the next several years he saw, by his own account, 1,500 bulls killed, enthusiastically involved himself in the world of matadors and promoters, and read widely in the exotic literature of the bullring.

In the fall of 1929, while *A Farewell to Arms* was on the best-seller lists, he turned out an article on bullfighting for Henry Luce's new magazine *Fortune*. Given the magazine's business focus, he kept largely to a matter-of-fact account of the economics of bull breeding and ring promotion, telling Max Perkins that he was writing in "journalese full of statistics" and keeping the article as dull as possible. He also informed the editor that there was potential in the subject for something better, for "every aspect I touch on if I could go on and write about would make a long chapter in a book."[11]

As soon as the article was finished he began work on the book, ending a gestation period of a half-dozen years. Composition went slowly, in part because of a litany of physical problems that included a severely fractured arm suffered in a car accident in Montana that required seven painful weeks in a hospital and a long rehabilitation period. When the book was finished there was a struggle with Perkins about the number of photographs it would carry, with Hemingway holding out for 112 as the "irreducible minimum" and settling for sixty-four together with a striking color frontispiece of *The Torero* by Juan Gris.[12] The book appeared finally on September 23, 1932, with a large first printing of 10,300 copies. Sales were brisk at first, then the Depression, the unusual subject, and bad reviews—the first Hemingway had experienced—took their toll. Hemingway later complained to one of his sons that *Ferdinand the Bull* made ten times more money.

The review that drew most attention was Max Eastman's attack in *The New Republic* some seven months after publication. Entitled "Bull in the Afternoon," it reduced Hemingway's fascination with the bloodletting of the bullring to overcompensation for doubts about his manhood. The most influential view, however, came from Edmund Wilson in a commentary on *Green Hills of Africa* in the same magazine three years later—a view that has shaped the critical response to *Death in the Afternoon* and indeed all of Hemingway's nonfiction ever since. Wilson, whose praise of the early stories helped launch Hemingway's career, held that when Hemingway wrote in his own voice in his nonfiction, as Hemingway, as against in the voice of the fiction, "the results were unexpected and disconcerting." Floundering excess replaced the admirable restraint of the fiction; everything became overdone, overblown. *Death in the Afternoon* had possessed some value as a treatise on bullfighting, Wilson allowed, but it still seemed an excessively heightened and even hysterical performance. *Green Hills of Africa* had no discernible saving grace; the account of African hunting seemed all "Hemingway maudlin." Wilson's overall judgment of both books was that "as soon as Hemingway begins speaking in the first person, he seems to lose his bearings, not merely as a critic of life, but even as a craftsman."[13]

Wilson was obviously right in noting the change in voice from the fiction to the nonfiction. There is even a change from the expected voice in a work of nonfiction. In *Death in the Afternoon* and his other fact books Hemingway not only writes in the first person but he makes explicit, often noisily and aggressively, what is ordinarily implicit in nonfiction, the writer presenting the material to the reader. And the effects of the changed voice are clearly different. The "admirable miniaturist in prose," as Wilson called Hemingway the fictionist, is replaced by the dominating personality, prickly and ironic, occasionally hectoring and long-winded.[14] The taut prose and implied meanings of the early fiction give way to loose language and frequent overstatement; the calculated omissions of the fiction are replaced with a rush of detail that seems meant to overwhelm the reader with an entire iceberg rather than its dazzling tip.

The sharply different manner of the nonfiction is cause for regret only if measured against exclusive admiration for the fiction. Taken by itself it delights, giving the nonfiction its essential character and adding an enriching diversity to the Hemingway canon. In the fiction Hemingway is guarded, secretive, suppressing more than he reveals, alert to Jake Barnes's remark that "You'll lose it if you talk about it." In the nonfiction the distinctive presence that has intrigued us so long leaps from the mediating shadows of the fiction, offering itself—usually *insisting* upon itself—to the reader directly.[15] As Lincoln Kirstein saw in a review of *Death in the Afternoon,* the book is "of all personal books, the most personal. The quality of its author's character is imprinted in the ink of the type on every page."[16] The same is true in varying degrees of all of Hemingway's nonfiction books: they stand out even among personal books as emphatically personal. Rather than being the centrally destructive element, Hemingway's presence in his books as Hemingway, together with an accompanying preoccupation with subject, is a central strength in each of them, infusing them with the quality that Walter Pater said was the very definition of art: life seen through a temperament.

This does not mean that Hemingway's nonfiction matches or surpasses the fiction in artistic merit or reader appeal. It simply means that the nonfiction, removed from the commanding presence of the fiction and considered by itself, offers its own rewards. Nonfiction writing—principally the book-length efforts, the heart of his achievement—occupied a considerable part of Hemingway's career and merits close and undivided attention. Although one of his biographers, Jeffrey Meyers, has claimed him as the most important American novelist of the century, his finest work was done in the short story; it was, as has often been remarked, his suited genre. Nonfiction writing in a vigorously personal vein and mingling fact and fiction was another. Few American fiction writers of major status have also produced such a varied and enduring body of nonfiction. As Hemingway himself might have put it in a combative mood, against the competition—James and Twain at the turn of

the century, Dos Passos, Agee, Hersey, Mailer, and Didion since—
he more than holds his own.

2

In a letter in 1926, Hemingway gave Max Perkins a foretaste of the
bullfight book, while at the same time providing a revealing statement of the complex ambitions of the work. He said the book was
"not to be just a history and text book or apologia for bull fighting"; instead, it was to be, "if possible, bull fighting its-self." It was
to have "all the dope" yet at the same time capture the depth of
feeling associated with the three-stage tragedy of the ring.[17] In the
bibliographical note at the end of the book Hemingway made the
same point with the flat remark that his aim was to explain bullfighting "both emotionally and practically."[18]

To achieve this end he constructed an expository account meant
to work on two levels—as he told Perkins in the letter, an account
with "outside" and "inside" elements.[19] The outside approach
would establish his presence in the book and his history as a bullfight aficionado; then, with the reader drawn into the subject
through identification with the writer's interest, he would go inside
the material in the sense of providing detailed sections on each of
bullfighting's aspects. The dual approach roughly describes the
structure of the book Hemingway produced, one that begins with
the directly personal in the opening sentence ("At the first bullfight I ever went to I expected to be horrified and perhaps sickened
by what I had been told would happen to the horses") and then
goes on to a succession of instructional chapters. More important,
the emphasis on outer and inner elements describes a dual perspective within the work that lifts it beyond the flat tones of a bullfight
handbook.

F. O. Mathiessen argued in *American Renaissance* that what
separated *Walden* from the simple records of experience produced
by Dana and Parkman was a doubleness in Thoreau "that made
him both participant and spectator in any event," capable of a
detachment that allowed him to stand apart from his experience.[20]
Although Hemingway professed not to be an admirer of Thoreau,

claiming he preferred his natural history free of literary trimmings, his method in *Death in the Afternoon* bears resemblance to Thoreau's. His typical doubleness is apparent in his capacity to give himself to the flow of fact about bullfighting while remaining apart from fact through insistence on his own commenting, reacting, feeling presence in the account. He retains an outer and inner perspective that keeps him from being one with his material, wholly given over to a factual account; fact is colored with personality, charged with personal reaction. This is exactly the approach Thoreau admired, that of a writer who, as he said, "was satisfied with giving an exact description of things as they appeared to him, and their effect on him."[21]

In Chapter 5 Hemingway pauses for a lecture on the tendency toward "erectile writing" in books about Spain (*DA,* 53). In such work the effort to portray feeling loses touch with actuality; it is "journalism made literature" through the "injection of a false epic quality" (*DA,* 54). The particular work he had in mind, and wished to dismantle, was Waldo Frank's *Virgin Spain,* published six years earlier and a book his was likely to be compared with. He was also describing through contrast his aims in *Death in the Afternoon,* a book that would also centrally concern itself with feeling, but true feeling in that it would remain tied to fact, moderated and contained by real knowledge. It would be journalism made literature not through applied lyricism but through depth and clarity of information and exactness of accompanying feeling. Hemingway offers the chaste simplicity of the Prado in Madrid as an analogy, a museum in which great paintings are hung with no attempt to create theatrical effect, emotion coming from observation of the work itself. Similarly, if feeling is crucial in his book, it is also crucial that it derive from fact and be controlled by fact.

Some of Hemingway's finest descriptive writing is found in Chapters 18 and 19 on the use of the muleta and proper killing in the bullring. The writing here is clear and exact yet there is always the close counterpoint of emotional reaction, especially when he considers the skills of such legendary matadors as Belmonte and Joselito. He is thinking of Joselito when he says that the heart of bullfighting's emotional appeal is the feeling of immortality it provides both bullfighter and spectator:

He is performing a work of art and he is playing with death, bringing it closer, closer, closer, to himself, a death that you know is in the horns because you have the canvas-covered bodies of the horses on the sand to prove it. He gives the feeling of his immortality, and, as you watch it, it becomes yours. Then when it belongs to both of you, he proves it with the sword. (*DA*, 213)

Chapter 19 ends with a flat account of the likely future of Spanish bullfighting, and perhaps as a reaction to the matter-of-fact ending of the section Hemingway turns in the final chapter of the book to a lyric account of Spain, "a country you love very much," as a way of returning to the counterpoint of personal response, again interweaving fact with feeling, the emotional and the practical (*DA*, 277).

Chapter 20 uses a device Hemingway would employ again—that of writing about what he had failed to write about, listing the evocative details of place and character that, if exactly captured, would "make all that come true again" (*DA*, 272). The chapter, admired even by readers who admire nothing else about the book, brings an emotional, sensuous response back to the center of interest and establishes Spain as the overriding subject despite the immediate focus on bullfighting. Spain is evoked as his last good country, replacing northern Michigan and Italy in his affection, but not even Spain is free of change, and there is also the unhappy fact that "we are older." Change itself occupies the final paragraph of the book, Hemingway maintaining, perhaps with more hope than conviction, that "I know things change now and I do not care" (*DA*, 278). Amid flux, all that endures is his work and trying to do it well, which brings him around to the book at hand and a final movement of the counterpoint of fact and feeling. Although he may not have gotten all of his affection for Spain into the book, he believes he has at least managed to say "a few practical things" about the country and the art of the bullring. From its outside opening the book comes, in its final line, to an inside close.

3

"Hemingway's tragedy as an artist," Cyril Connolly once said, "is that he has not had the versatility to run away fast enough from his imitators. . . . A Picasso would have done something different; Hemingway could only indulge in invective against his critics—and do it again."[22] The charge, if it has merit at all, applies more to the fiction than the nonfiction. Hemingway's fact books are each remarkably different. Each has its roots in a conventional genre of nonfiction writing—handbook, travel book, journal, reportage, memoir—and each (with the exception, as noted later, of the unfinished African book) evolves into highly individual work marked by a good deal of technical ambition and imprinted with personality. A familiar genre was a starting point; the end was all his own.

Green Hills of Africa is the most experimental of Hemingway's nonfiction books. Early in 1934, shortly after returning from a three-month African safari with his second wife, Pauline, and a Key West friend, Charles Thompson, he began writing a story about the experience that soon turned into a factual account in which he reconstructed events in a fictional manner. The experiment, he announced in the book's foreword, was to see whether an "absolutely true book" devoted to re-creating the shape of a country and a month's action in that country could compete with a work of the imagination in narrative power. Although the finished work was hardly the purely truthful work he claimed, he attempted to stay generally within the facts as recalled while the events and emotions of the safari were still fresh in mind.

As with the bullfight book, *Green Hills of Africa* was in some measure meant to fill a literary void, although it was a book lacking the instructional purpose of the earlier book. Its ambitions were more wholly on the plane of feeling—to re-create the sights of Africa and the emotions it evoked through concentration on the narrative of the hunt. It was a book for the reader to live inside of rather than consult for information, and so it was more nearly the "enough of a book" Hemingway had limned in the final chapter of *Death in the Afternoon,* a book to make experience come true again (*DA,* 270). In the African book he tells of reading a story of Tolstoy's during an interlude in the hunt and feeling as if he is "living in that Russia again." *Green Hills of Africa* similarly attempts to draw the reader into the work through description and narration, experiencing the place and action that Hemingway had experienced. It is meant to enchant in exactly the same manner as the Tolstoy story, allowing the reader to see "the river that the

Tartars came across when raiding, and the drunken old hunter and the girl and how it was then in the different seasons."[23]

In the book the white hunter, Pop (the name Hemingway gives the safari guide, Philip Percival) and Hemingway agree that they have read nothing about Africa that captures the way they feel about it. For them Africa is still virgin literary territory. Hemingway declares that he would "like to try to write something about the country and the animals and what it's like to some one who knows nothing about it," and Pop replies: "Have a try at it. Can't do any harm" (*GH,* 194). Hemingway had earlier pointed out that without speaking the native languages you cannot talk with people and overhear conversations, leaving you unable to get anything that's of anything but journalistic value." Because of this limitation, he informs Pop that "if I ever write anything about this it will just be landscape painting until I know something about it" (*GH,* 193).

The book finished, Hemingway wrote to Max Perkins that when he began writing that was indeed all he wanted to do—make the country come alive again as landscape painting, capturing what he had seen and felt in Africa. It was all he could write—a travel book of journalistic value—given the brevity of his stay and lack of languages. He went on to inform the editor that in the process of writing he had discovered a "story"—not an invented story, he insisted, but the actual things that had happened—and this in turn had provided the work with structure and dialogue and action. It was no longer simply an account of his first seeing of the country, something to get stated and of likely importance only to himself. It was also "that wonderful goddamned Kudu hunt—the relations between the people—and the way it all worked up to a climax." With the discovery of a story within the experience the work now seemed to him maybe the best thing he had ever written since it was all true yet far removed from mere journalistic value. "True narrative," he instructed Perkins, "that is exciting and still is literature is very rare."[24]

Perkins had told Hemingway that nobody could map out the organization of *Death in the Afternoon,* a book that gave the editor an impression of growth more than plan. *Green Hills of Africa* was the opposite. It had a carefully worked out, complex structure directed to a narrative center, the down-to-the-wire hunt for the elusive kudu antelope. Hemingway chose to concentrate attention on a period of the safari in which he and Karl (the name given Charles Thompson) are hunting in upland country and on foot before the coming of the rainy season that will bring the adventure to an end. The book opens just before the close of the period, on the tenth day of failed hunting for a bull kudu and with three days left for escape to the coast ahead of the rains. A chance meeting in the bush with an Austrian who remembers some of Hemingway's poems in a German periodical leads to a literary discussion in which Hemingway becomes a self-appointed lecturer on American writing, giving the section its title, "Pursuit as Conversation." The second section, "Pursuit Remembered," turns back to an earlier time in the safari, re-creating hunts for lion and rhino as well as kudo and establishing the edgy shooting competition between Karl and Hemingway.

With part three, the story returns to the time of the opening scene and records repeated failures in the kudu hunt. In this section, "Pursuit and Failure," the talk turns to literary matters again, with Pop and Hemingway mentioning their disappointment with African books and Hemingway describing the book he would like to write. The hunt for kudu now looks "washed up" to him, and the safari is about to end on a dismal note (*GH,* 206). At this late point news comes of a place, three or four hours away by car, abounding in kudu. A plan is hastily formulated in which Hemingway will hunt the new country alone while Pop and Pauline break the main camp and join later with Hemingway and Karl.

With part four, "Pursuit as Happiness," Hemingway and a native entourage enter "the loveliest country of Africa," an unhunted pocket in the great sweep of the continent (*GH,* 217). After a camp is made Hemingway begins the hunt with only an hour left of daylight. Within minutes he has killed two bull kudu, and after the skins and heads are back in camp there is a scene of joyful release and comradeship, with Hemingway drinking beer and roasting meat with the natives and stumbling through conversation without a common language. When he eventually rejoins Pop and Pauline he finds that Karl has taken a kudu with larger horns and he is suddenly "poisoned with envy" (*GH,* 291). But in the morning his dark feelings are gone; his two kudu look good enough and he is at ease with Karl.

He asks Pop about the ritual of pulling thumbs he had gone through with the natives after killing the kudu, learning it is a sign of brotherhood. "You must be an old timer out there," Pop explains (*GH*, 294). The final triumph is not just the two kudu, but also Hemingway's admission to a deeper level of African experience, a victory beyond the reach of Karl's streak of good luck. He has passed beyond the fleeting experience of the hunter on safari into a deeper union with the land and its inhabitants. The book he will write—casually mentioned in the postscript as something to keep alive Pauline's memory of Pop—will be more than he first anticipated ("something about the country and the animals and what it's like to some one who knows nothing about it"). The successful pursuit of the kudu, wresting happiness from failure, had deepened the safari and the book about it, providing a "story" and with it the possibility of a true narrative that is exciting yet possessed of the permanence of literature.

In addition to Edmund Wilson's aversion to Hemingway's dominating presence in the book, a common complaint addressed to *Green Hills of Africa* is that Hemingway overvalued the material of a month's hunting action, the work finally lacking sufficient hold on the reader's interest. The novelist William Kennedy put this view bluntly, maintaining that the book "perished in the bush from overkill: too much hunting detail, too much bang-bang banality, insufficient story."[25] Certainly the story is confined to the enclosed world of the safari, the repetitious daily round of hunting with the dawn, rest in the afternoon heat, hunting again until dark, and talk around the campfire, with a change coming only with the movement to new country and new pursuit. From one point of view, there is not much to it.

For Hemingway, however, the days are highly charged and of absorbing interest; he wants only to lengthen them out, then repeat them in the future. The attempt is to draw the reader into sharing the exhilirating preoccupation with a sharply circumscribed place and action, and leave all else blotted out for the moment. The story line he found in the experience ("the way it all worked up to a climax") seemed to Hemingway a compelling element that assured the reader's interest, but the real meat of the book is the reconstruction of riveting physical and emotional involvement with the hunt.

In a letter to the Russian critic Ivan Kashkin just after the publication of *Green Hills of Africa,* Hemingway drew a distinction (as he had in the book in dialogue with the Austrian) between the joys of writing and action. "The minute I stop writing for a month or two months," he said about the latter, "and am on a trip I feel absolutely animally happy."[26] On this level—the account of an action that had stimulated in him an animal happiness—the book is as engrossing as Hemingway thought it was. His African landscape painting, the intricate structure he worked out, and the drama of the kudu hunt add to the pleasure, but it is the portrayal of total, mesmerizing absorption in physical activity that gives the book its special flavor.

4

After a concentrated period of fact writing that had produced two major books but little enthusiasm, Hemingway plunged back to fiction, drawing on Havana and Key West material for *To Have and Have Not* and the safari for "The Snows of Kilimanjaro" and "The Short Happy Life of Francis Macomber." Nearly twenty years would pass before he ventured on another sustained period of nonfiction.

With *Death in the Afternoon* and *Green Hills of Africa* he had been young, vibrant, and successful. When he returned to extended nonfiction in the final decade of his life, his physical and mental health was in precipitous decline, there had been long dry periods of writing, and though his commercial stock had soared his critical reputation had been severely battered. He liked to describe himself now as a crafty old lion in the African veldt who knew his limits but could still transcend them when he wanted. The chosen field of effort at the end was largely in nonfiction where he summoned energy for three book-length works, each drawn from nostalgic returns to the locales of past exploits: Africa, Spain, Paris.

But effort was often more evident than cunning in his final decade. Two of the accounts were swollen and frequently flat performances, lacking the quality of his earlier Spanish and African books; he left one unfinished and both unpublished. Although the Paris sketches that came to be called *A Moveable Feast* inspired a warm wave of critical applause after the book's posthumous publication, the work as a whole provided only occasional echoes of his best undertakings in nonfiction. In the end the old lion turned once again to the fact writing with which he had begun his career, and with steadfast effort, but he could rise above the burden of his last years only fitfully and with little sense of satisfaction.

The return to Africa in August 1953 was an attempt both to recapture old joys and to make a fresh start, desires that had been attributed to the writer, Harry, in "The Snows of Kilimanjaro": "Africa was where he had been happiest in the good time of his life, so he had come out here to start again."[27] The best part of the second safari came when the hunting was largely over and Hemingway and his fourth wife, Mary, were alone and simply observing the animals. An added pleasure was an appointment as an honorary game warden, an assignment Hemingway handled with a mixture of seriousness and high spirits, touring shambas on patrols, fielding a steady stream of native complaints, dispatching marauding animals. The safari's ending, however, turned into a nightmare of bad luck when a plane carrying the Hemingways crashed on a sightseeing flight and then a second plane crashed while taking them for medical treatment.

Back at his home near Havana in the summer of 1954, Hemingway began a discursive, mostly factual day-to-day journal of the safari. The work was soon interrupted by the announcement in October of his Nobel Prize, but by the end of the following October he had managed to bring the manuscript to 650 typed pages. On February 27, 1956, he abruptly stopped work in the middle of a sentence on the 850th page, wrapped the manuscript in cellophane, and flew off to Peru for the filming of fishing sequences for *The Old Man and the Sea.* Apparently, the African book was never worked on again. In the fall, Hemingway was in Spain for the bullfights. He then traveled with Mary to Paris where, in the Ritz Hotel, they discovered two trunks left behind and forgotten when Hemingway had moved to Key West in 1928. In the trunks were notebooks and clippings and sheaves of typed fiction. According to Mary, Hemingway sat for hours on the floor beside the trunks reading his early work. When he returned to Cuba the Paris material went along with him together with the first stirrings of a new kind of fact book.

The African book remained hidden away until the winter of 1971–1972 when *Sports Illustrated* magazine published three lengthy excerpts. The manuscript's existence had come to light in a 1969 inventory of Hemingway materials in which it was described as an unfinished "autobiographical account of duties as volunteer ranger at the Masai game preserve at foot of Mt. Kilimanjaro in late 1953."[28] The published excerpts totaled some 55,000 words, just over a quarter of the manuscript, and were edited by Ray Cave, an editor of the magazine, in such a way that all deletions were indicated with ellipsis marks, asterisks, and space breaks. The full manuscript is now among the Hemingway papers at the John Fitzgerald Kennedy Library in Boston but has yet to be opened for inspection.

The "African Journal," which is the title Cave gave the untitled manuscript, exhibits Hemingway's usual mixture of fact and invention—a work of nonfiction only in the sense of his decision to stay more generally within the facts than he typically did in his fiction. Mary, who was in a position to know, identified the work as "semifictional," yet she also paid tribute to the "tape recorder in his head" that enabled Hemingway, who had made no notes in Africa, to capture the safari with considerable accuracy.[29] "I have checked my diary to verify things," she told Cave, "and he was consistently correct. He might embellish a scene, but he did that with everything in life."[30]

Judging from the magazine selections, Carlos Baker's assessment of the entire African manuscript in his biography of Hemingway as "almost completely formless, filled with scenes that ranged from the fairly effective to the banal" is probably correct.[31] Cave also found the work as a whole unstructured and undisciplined, and considered it no more than a first draft. The personal voice, clearly Hemingway, is there again in the nonfiction, though not as dominating and authoritative—as *present*—as before. And there is nothing of the carefully worked structure of *Green Hills of Africa*. For readers who find the hunting tedious in the account of the first safari, this time there is a minimum of step-by-step shooting detail. Mary's hunt for a lion and Hemingway's for a leopard, the central events of the magazine excerpts, are relatively uneventful affairs, lacking the intensity and drama of the earlier hunts for rhino and kudu. There are some striking passages in the work—in one, dry cider and pillows stuffed with balsam needles suddenly remind Hemingway of the tastes and smells of Michigan when he was a boy—but for long stretches the writing is as limp and uninspired as the opening sentence of the first magazine installment: "Things were not too simple in this safari because things had changed very much in East Africa."[32]

Given the unfinished state and apparent quality of the work, it is probably wise that the African manuscript has not yet been put into book form. Yet in her 1976 autobiography, *How It Was,* Mary listed it among the Hemingway manuscripts awaiting editing and publication, and given the fate of *Islands in the Stream* and *The Garden of Eden,* unfinished works that eventually came out in highly edited editions, it seems likely that a book version will one day appear. Until then, we can only glimpse the kind of book Hemingway may have had in mind in the magazine installments.

In *Green Hills of Africa* he told of his desire for a second safari that would be conducted with leisure, providing a new experience of Africa. There would be no hunting pressure and no need to accumulate trophies. This time, he would live in the country observing the animals and writing when he chose. He would

> get to know it as I knew the country around the lake where we were brought up. I'd see buffalo feeding where they lived, and when the elephants came through the hills we would see them and watch them breaking branches and not have to shoot, and I would lie in the fallen leaves and watch the kudu feed out and never fire a shot. . . . I'd lie behind a rock and watch them on the hillside and see them long enough so they belonged to me forever. (*GH,* 282)

The honorary game warden's position gave him the altered relationship to Africa that he wanted, one in which he to some extent settled in, providing he and Mary, as he put it in the manuscript, with "this great privilege of getting to know and live in a wonderful part of the country and have some work to do that justified our presence there."[33] That work, such as it was, might have provided a firm enough spine for a fresh African book, just as the kudu hunt had in the first book and as the instructional, handbook mode had in *Death in the Afternoon.* Consigning the manuscript to cellophane oblivion, perhaps an appropriate act in a critical sense, seems an unfortunate one given the African sequel he might have produced.

5

Hemingway broke off work on the sketches that would become *A Moveable Feast* for a season of bullfights in Spain in the spring of 1959. The two leading matadors of the day, Antonio Ordóñez and Luis Miguel Dominguín, were about to engage in a series of *mano a mano* contests (each bullfighter killing three bulls rather than the usual bullfight card of three matadors and two bulls each) that he was determined to see. He had also hatched another writing task for himself, one that provided a professional excuse for the return to Spain. Scribner's was planning to reissue *Death in the Afternoon* and he wanted to update it by adding an appendix dealing with the current bullfight scene, an idea he had long had in mind.

Life magazine learned of the project and made him an attractive offer for prior magazine publication. The agreement called for a 5,000-word news story. The piece was to appear shortly after the bullfight season, with the focus on what it was like for him to return to Spain and on the summer's rivalry between Ordóñez and Dominguín. It must have struck Hemingway as an undemanding task, and one that avoided the ephemeral nature of magazine journalism since the material would eventually be recycled in the reissue of *Death in the Afternoon*.

With an entourage that included Bill Davis, whose home near Málaga provided a base for Hemingway and his wife, and his friend A. E. Hotchner, he hurtled around Spain on an arduous schedule of bullfights and accompanying revels. His health still showed the effects of the African plane crashes and there were increasing signs of mental instability. In October, after Mary left Spain to ready their houses in Cuba and Ketchum, Idaho, but also to escape her husband's erratic behavior, he stayed on to work on the *Life* article, beginning it on October 10 with an account of his return to Spain in 1953, the first time he had been back in the country since Franco's victory in the civil war. He quickly finished 5,000 words but kept on writing. As described by the Spanish journalist José Luis Castillo-Puche, who was close to Hemingway at the time, the work lurched forward amid confusion and anxiety:

> I had seen him . . . putting sheet after sheet of paper away in his files and then taking them all out again, pages only half finished,

pages so full of corrections that they were scarcely legible. For the first time since I had known him, I saw him get all confused, tear up whole sections of his manuscript. . . . For the first time in his life Ernesto had made a mess of what he was writing.[34]

Back in the less hectic settings of Cuba and Ketchum the manuscript kept growing and a new idea formed, that the work might appear as a separate volume rather than part of a revised edition of *Death in the Afternoon*. By the end of May he had a first draft of what he called "this about the bulls that comes after Death in the Afternoon."[35] Earlier he had put off *Life*'s editors by telling them he had discovered a different kind of story that required greater length, "the gradual destruction of one person by another with all the things that led up to it and made it." So it had been necessary to establish the different personalities of the two matadors and their different approaches to the art of the bullring. Another factor leading to increased length was his need to make something of lasting quality after the Ordóñez-Dominguín duel was no longer newsworthy. As he now conceived it, the report was to "have some unity and be more than the simple account of the mano a manos which were no longer news and had been picked over by various vultures and large bellied crows."[36]

When time came to present the manuscript to *Life,* Hemingway called Hotchner to Cuba for editing help. As Hotchner recalled, the manuscript he dealt with had 688 typed pages, 108,746 words. The two of them carved out a manuscript of 53,830 words that Hotchner then carried off to New York.[37] The first of three lengthy installments appeared in *Life* on September 5, 1960, the white-bearded author featured on the cover and with an accompanying note indicating that a complete edition of *The Dangerous Summer* would be published by Scribner's the following year. But it was not until 1985, a quarter-century after Hemingway's death, that the book appeared in a sharply reduced version of about 45,000 words that had been edited from the *Life* manuscript by a Scribner's editor, Michael Pietsch.

The book is not only an edited version of Hemingway's complete manuscript but a far smoother, more polished, and less detailed work. The book version has an orderly structure of chapters, whereas the manuscript has no internal divisions, and the book

removes detail on bullfighters other than Ordóñez and Dominguín. The major difference is that the book follows the manuscript only through its first 493 pages, eliminating the last 195 pages
of the manuscript and in so doing removing its personal and som-
ber conclusion. The manuscript ends with Hemingway back in
Cuba, writing, and his mood decidedly down. A new bullfight
season has begun in Spain and he halfheartedly feels he ought to
attend to check on new developments, but he concludes that his
work about the past season is finished, his "monument" to it
made.[38] If not, he adds in a melancholy and wholly out-of-
character remark, he hopes that someone else can carry the project forward from where he ends.

In a passage appearing in the book version—although not in the
Life series—Hemingway says that before coming to Spain he wrote
Ordóñez that he

> wanted to come over and write the truth, the absolute truth, about
> his work and his place in bullfighting so there would be a permanent
> record; something that would last when we were both gone.[39]

How to craft such a lasting record from the relatively thin journalistic material of a summer's spectacle was the writing task he set
himself. Part of his approach is simply the piling up of bullfight
detail, an attempt to create permanence through density—an easy
target, as it turned out, for the editorial pencil. More important is
the personal stamp Hemingway gives the report, linking it to the
rest of the nonfiction even though his distinctive presence is less
dominant and intrusive than in either *Death in the Afternoon* or
Green Hills of Africa. He once again locates himself at the center
of attention, the two matadors seeming to perform for him, awaiting the implacable judgment that will consign them to their places
in bullfight history. On the deepest level, Hemingway's domina-
tion of the material involves transforming it into a dramatic account of familiar Hemingwayesque dimension, one that may have
had little to do with the reality of events. With the second bullfighting book he found or created, as he had with kudu hunting in the
first African book, the imaginative story meant to give life to
events.

Although Hemingway professes to be absolutely just in his ap-
praisals of the two bullfighters and to find the rivalry a terrible

strain on his friendship with both, it is clear from the beginning of the book that he is drawn emotionally and critically to Ordóñez. "I could tell he was great from the first long slow pass he made with the cape," he says of his first sight of the younger bullfighter in action (*DS,* 49–50). Dominguín is admired as a cynical, worldly-wise companion and for his grace and skill in the ring; as a banderillero he is unsurpassed. Hemingway is not moved by his work with the cape, and he later discovers that Dominguín has a penchant for crowd-pleasing tricks. In the summer of 1959 the two are locked in fierce rivalry, with Dominguín the more successful matador and the better paid, and Ordóñez out to prove he is the superior figure.

The climax comes in Bilbao, before the most severe of bullfight audiences, when Dominguín is seriously gored, a horn entering his groin and penetrating to the abdomen. Ordóñez, alone now in the limelight, kills his last bull in *recibiendo* fashion—standing motion-less with the sword and receiving the bull's charge—a classic, dan-gerous, and nearly forgotten way of killing that Hemingway had celebrated in *Death in the Afternoon* as "the most arrogant dealing of death and . . . one of the finest things you can see in bullfight-ing" (*DA,* 238). He said in that book that he had seen it properly completed only four times in more than 1,500 kills; with Ordóñez he sees it again. In the summer of 1956, Ordóñez had killed bulls *recibiendo* as well, and in one instance he had dedicated a bull to Hemingway with the words, "Ernesto, you and I know that this animal is worthless but let's see if I can kill him the way you like it" (*DS,* 58). Now, at the end of the summer of contests with Dom-inguín, he repeats the act, convincing the audience of his preemi-nence and turning the book into at least an approximation of an appendix to *Death in the Afternoon,* the original plan that had stimulated Hemingway's renewed study of courage and death in the bullring.

Considered simply as a work of reporting, *The Dangerous Sum-mer* raises several problems. Spanish observers have pointed out that Ordóñez and Dominguín did not have a true rivalry in the sense of matadors of equal standing in competition for a number of years and involving many *mano a mano* encounters. Hemingway's presence turned the attenuated duel into a media event, and his eventual account of it thus became—in one commentator's view—

"an excessive monument to a not-so-great event."[40] Observers have also taken Hemingway to task for his overwhelming bias in favor of Ordóñez, arguing that Dominguín was not as outclassed as Hemingway maintains nor destroyed as a major *torero* as result of the head-to-head contests. Dominguín himself stoutly resisted Hemingway's characterization, claiming that the writer had deliberately heightened his rivalry with Ordóñez into deadly competition.

Yet, a rigorous journalistic effort to capture what he called the "absolute truth" was something for which Hemingway had little inclination. His journalism was always a form of antijournalism, emphasizing his own reacting presence and mingling observation with invention, always making more than describing. He sought to shape what he observed and experienced into an account of universal value, something that would hold up as the truth of art long after the truth of events had faded from memory. Ordóñez' method of killing at the end of the climactic contest in Bilbao links him with the eighteenth-century matador Pedro Romero, who regularly killed *recibiendo* and who had given his name to the flawless bullfighter in *The Sun Also Rises* (a figure in turn modeled on Ordóñez's father, the bullfighter Cayetano Ordóñez), and with the indomitable Maera whom Hemingway had written about in glowing fashion in "The Undefeated" and *Death in the Afternoon*. With Antonio Ordóñez, Hemingway re-created his idealized matadors, blending what he saw with what he wished to believe, coming around full circle to one of the imaginative centers of his work, the man of physical conduct worthy of unqualified admiration.

6

According to Mary, after the manuscript of *The Dangerous Summer* was finished and the *Life* installments had appeared Hemingway was so repelled by the venality of the bullfight business that he put book publication aside and returned to his Paris sketches. After his death in July, 1961, she found the typed manuscript in his workroom in Ketchum together with a draft preface and a list of titles, including *A Moveable Feast*. Since picking a title was among the last things he did before publication, she concluded that he considered the book in final form.

Mary and an editor at Scribner's, Harry Brague, saw the manuscript through publication. When the book appeared on May 5,
1964, the first of Hemingway's posthumous works, there was no
mention of any editorial change, but there was a note by Mary
saying only that Hemingway finished the book in Cuba in the
spring of 1960 and made some revisions in Ketchum the following
fall. We now know that the book departs significantly from Hemingway's manuscript, so much so that some critics believe a new
text is necessary if we are to read the book Hemingway actually
left behind.[41] Yet whatever its inadequacies as the work he intended, the published version of *A Moveable Feast,* like that of *The
Dangerous Summer,* is the one at hand for the time being. It is the
most widely read and widely admired of Hemingway's nonfiction
books—the book that has been seen as the fitting capstone of his
career.

The appearance of Gertrude Stein's *The Autobiography of Alice
B. Toklas* in 1933 got Hemingway, with revenge in his eye over the
charge that he was yellow, talking about another man-of-letters
endeavor: memoirs. In an article in 1924 he had tossed off the
remark that "it is only when you can no longer believe in your own
exploits that you write your memoirs."[42] Now he told the writer
Janet Flanner that "by jeesus will write my own memoirs sometime
when I can't write anything else. And they will be funny and accurate and not out to prove a bloody thing."[43] Max Perkins received
the same boast. After mentioning reading an installment of Gertrude Stein's book in the *Atlantic Monthly,* Hemingway promised
the editor that he was "going to write damned good memoirs when
I write them because I'm jealous of no one, have a rat trap memory and the documents."[44]

As a subject that especially needed straightening out, the old
days in Paris were always high on his list. In "The Snows of Kilimanjaro" Harry evokes the Paris quarter in which "he had written
the start of all he was to do" but that he had never captured in his
work: "No, he had never written about Paris. Not the Paris that he
cared about."[45] In the African manuscript Paris makes a passing
appearance as a city known and loved so well that "I never liked to
talk about it except with people from the old days."[46] What apparently got Hemingway writing about it was the discovery in 1956 of
the trunks in the Ritz Hotel.[47] He began work the following year in

Cuba with an account of his first meeting with Scott Fitzgerald that he was thinking of sending to the *Atlantic Monthly,* which had asked him for a contribution to its centenary issue. Soon he turned to other remembrances that he called sketches, a sardonic reference to the way his early stories had been characterized by uninterested editors.

Mary, who typed the sketches as he finished them, recalled that she found them disappointing. "It's not much about you," she objected. "I thought it was going to be autobiography." Hemingway replied that he was writing "biography by *remate*"—a jai alai term Mary interpreted as meaning that he was trying to create a portrait of himself by reflection; his life seen in light of others.[48] As a memoirist, Hemingway was indeed working with considerable indirection, moving freely across a time frame of the years 1921–1926, establishing only a loose continuity among the sketches, leaving more out than he put in. Nonetheless, the discrete, usually brief sketches, ordinarily opening with a generalized page or two about a place or activity and coming around to an anecdotal ending, were an inspired form at this point, and the book grew rapidly.

By the summer of 1958 he had eighteen sketches behind him and was telling a friend that he had written "a book, very good, about early earliest days in Paris, Austria, etc.—the true gen on what everyone has written about and no one knows but me."[49] In November of the following year he left the manuscript with his publisher during a stopover in New York, with instructions that it be returned to him in Ketchum for some final work. According to A. E. Hotchner's account of the final year and a half of Hemingway's life, little was ever done to the manuscript again; long hours were spent poring over the pages, but there was no real writing. Hotchner remembered his bitter complaints:

> I've got it all and I know what I want it to be but I can't get it down. . . . I *can't.* I've been at this goddamn worktable all day, standing here all day, all I've got to get is this one thing, maybe only a sentence, maybe more, I don't know, and I can't get it. Not any of it. You understand, I *can't.*[50]

With *A Moveable Feast* Hemingway turned to a different nonfiction genre, a memoir approached through a fragile mosaic of twenty self-enclosed sketches. The personal voice and powerful personality are present once again, though the voice is muted and wistfully reflective now, resembling more the detached authorial voice of the fiction than the aggressively personal Hemingway of the earlier nonfiction that Edmund Wilson had found so offensive. The mixture of fact and fiction in the book resembles the fiction as well in that it contains an even higher proportion of invention than usual, Hemingway reshaping his life to the end. But for all that, *A Moveable Feast* is not fiction. Hemingway clearly approached the book as a work of nonfiction drawn from recollection, though one bearing a remark in the preface (in anticipation, perhaps, of libel actions) that the book could be regarded as fiction if the reader wished and with his usual awareness, as he had said in *Death in the Afternoon,* that "memory, of course, is never true" (*DA,* 100).

One large aim in the book was to set the historical record straight about the old days and old acquaintances. He noted in a letter that "it is important that I should write about the Paris part as no-one knows the truth about it as I do and it is an interesting time in writing." The record-straightening aim was highlighted in the preface with the remark that the book might "throw some light on what has been written as fact."[51] It was to be a factual book, then, to revise what others had claimed as fact, yet a book free of illusions about recovering exact historical truth. It was a book relying on memory, but memory, as Hemingway well knew, always bore with it distortions, among them the burden of present concerns that colored the sense of the past. In *A Moveable Feast* he was not re-creating Paris as it had been but, in the pose of a memoirist this time, fashioning it through selection and invention into the way he chose now to remember it.

Rather than a thoroughgoing account of the Paris he had known, Hemingway provides a set of highly eclectic memories—memories always in the service of the self, with everything aimed inward. He recalls old friends and enemies, settles old scores, and retells old tales. All the stories were arranged to place himself in an attractive light, his own sterling "discipline," as he remarks at one point, set off against the "egotism and mental laziness" of his chosen cast of flawed figures.[52] Three sketches devoted to Gertrude Stein are directed with skilled vituperation to a passage of overheard conversation, Hemingway delivering in print the low blow he had long been saving up. Scott Fitzgerald is on the receiving end of the most extended recollection, one arranged with a heavy hand to show Hemingway's contrasting superiority.

The absurdist conclusion to the Fitzgerald recollection comes with Hemingway reassuring Fitzgerald in the "A Matter of Measurements" sketch that his penis is of adequate size. Fitzgerald remains doubtful even after a trip to the Louvre and the inspection of male statues. The scene breaks off with Fitzgerald leaving for the Ritz and Hemingway leaping ahead in time to a passage of dialogue after World War II with a bartender in the hotel who can not remember Fitzgerald. Georges recalls perfectly his first meeting with Hemingway but wonders, "Papa, who was this Monsieur Fitzgerald that everyone asks about?" (*MF,* 191). Hemingway's response is to pledge that he will write about Fitzgerald in a book about the early days in Paris, putting him in "exactly as I remember him the first time that I met him" (*MF,* 193). But the account in his book of remembering is hardly so innocent or so wholly concerned with the historical past. Hemingway was not so much looking back in his portrait of Fitzgerald as ahead, trying to fix the lasting literary fate of his great rival from the Paris days.

If establishing the record, Hemingway fashion, is one strain in *A Moveable Feast,* another is the portrait of the artist, in love with work and wife, at the beginning of the famous career. Readers have generally found this the more appealing side of the book, yet it is equally elliptical in its treatment of things past, glancing in its characterization, and devoted to refashioning the past in light of the present. Hemingway constructs a romance of writing in which all he needs is a warm café, a pencil, and a cheap notebook. It is a magic time in which stories seem to write themselves and he is immune to distractions. Even when things are not going well he simply admonishes himself to remember that

> "You have always written before and you will write now. All you have to do is write one true sentence. Write the truest sentence that you know."

And it would work. It was "easy then because there was always one true sentence that I knew or had seen or had heard someone say" (*MF,* 12). Because of the hard struggle of writing in the present that he had confessed to Hotchner (". . . I can't get it. Not any of it"), writing in the reconstructed past takes on all the more imagined ease. In *A Moveable Feast* all incapacity is lifted; there remains only the memory of good work and good luck:

> The blue-backed notebooks, the two pencils and the pencil sharpener (a pocket knife was too wasteful), the marble-topped tables, the smell of early morning, sweeping out and mopping, and luck were all you needed. (*MF,* 91)

As Hemingway chose to recall them, in those days "work could cure almost anything"—a doctrine, he insisted from the vantage point of the present, he still believed (*MF,* 21). The other cure was his love for his first wife, Hadley, imagined back into existence in terms as idyllic as his portrayal of the ease of writing, and perhaps for a similar reason—as a response to the severe strains in his present marriage to Mary. In the memoir Hadley is portrayed as the impossibly perfect mate for the young writer. She accepts their (much exaggerated) poverty without complaint; she is instantly ready to do what he wishes and go where he wishes—a dimly seen but always supportive presence whose sole purpose is to selflessly accomodate herself to her husband.

Yet the marriage to Hadley was not all that secure, and as the relationship is described, Hemingway, in a chastened voice coming from outside the time frame of the sketches, hints at the failure to come. On one occasion, after saying how lucky they were, he adds that he foolishly had not knocked on wood; in a sketch about the horse races at Auteuil and Enghien, he says directly that "racing never came between us, only people could do that" (*MF,* 61). The end is described in elaborately veiled fashion in the book's final sketch about two winters spent in Schruns in Austria, the opening line introducing the story of disruption to follow: "When there were the three of us instead of just the two. . . ." (*MF,* 197) Hemingway's affair with the third figure, Pauline, is temporarily put aside when he returns from a trip to New York to establish himself with Max Perkins and Scribner's, and the "lovely magic time" in the mountains with Hadley is resumed. With the return to Paris in the spring, however, "the other thing" is resumed as well, and the marriage, and with it the first chapter of his life in Paris, comes to an end. The city replaces the marriage in a lyric coda to the sketch as the place where, once, he and Hadley "were very poor and very happy" (*MF,* 210–11).

With the publication of *A Moveable Feast* the old lion seemed, miraculously, to draw a fresh masterpiece out of his nearly exhausted fund of physical and creative energy. One critic, for example, found the work "proof of the strength he could still muster," adding that "the book is new, and stands with the best of his early stories."[53] Set against a background of Hemingway's later work in both fiction and nonfiction, the book did seem engagingly fresh, in large part because it drew attention back to the vivid, economical world of the stories. The voice in some of the sketches strongly recalls the early work, but the older, embattled voice is there as well, speaking from the vantage point of present struggles, concerned to stay sound and good in his head until morning and the resumption of work.

The mingling of voices coming from different time frames gives *A Moveable Feast* its uneven tone, sometimes joyous with work and love and the delights of Paris, sometimes darkly melancholy and burdened with intimations of death. The weight of the present influences the book's structure as well as its tone. Overall design had been a problem with both the second African account and the report on the Ordóñez–Dominguín rivalry. Hemingway found a form with *A Moveable Feast* that could be sustained—the real triumph of the old lion's craft at the end. His greatest strength had always been for the short narrative and a treatment of people more directed to capturing the veneer of personality than internal realms; for his nonfiction at this point, the brief sketch—perhaps influenced by the discovery of the old material in the Paris trunks or the recollection that his early fiction had been dismissed as sketches—was an ideal form. It freed him from the need to build the account of Paris days into a connected whole, allowing him to rest with only a casual linking of the sketches and to slide over their differing aims and moods. It enabled him, finally, to practice in nonfiction something of the fictional discipline of omission that he recalled in the book as "my new theory," the aura of suggestion cast over the material that left the reader feeling more than was stated (*MF*, 75).

The practice of omission turned *A Moveable Feast* more toward the body of his fiction, especially the early stories, than the body of his nonfiction. As a result, when viewed in light of his major nonfiction efforts in *Death in the Afternoon* and *Green Hills of Africa* it can seem a minor work—thin, driven by cross purposes, too given over to surface observations in its portrayal of both the young Hemingway and his Paris contemporaries. But comparisons are beside the point. In the Paris book he again turned his hand to something new in nonfiction—to a form of autobiography pursued through the fragmented, roundabout means of reflection. Once again he started with a familiar genre of fact writing and once again shaped it, through the medium of fictionlike sketches, to his own special creation—a work suffused with temperament, the actual drawn deeply inside an imaginative construction, something made rather than recalled and so possessed of continuing life. *A Moveable Feast* stands alone, as do each of his other nonfiction books. There are no proper comparisons.

Hemingway seems to have intended a second volume of Paris memories, picking up with the story of his marriage to Pauline and his new life in the old city. In the preface to *A Moveable Feast,* after listing some things missing from the work, he remarks that "we will have to do without them for now," and Mary had mentioned after the work's publication that "he had planned, I guess, to do a second book about Paris."[54] Certainly there was more about the city that he might have written, and more about his adventurous life in many places that might have been spun out in a string of memoirs. There were more events that needed his clarifying "true gen," more rivals to be put in their places, and more wives to be recalled. He might even have gotten around to a fact book about fishing, a work Max Perkins had once brought up in a letter: "By the way, couldn't you write a wonderful book about fishing sometime, full of incidents about people, and about weather, and the way things look, and all that?"[55] Looking back at what he did accomplish in nonfiction, the regret is not that time was diverted from fiction but that there was not more time to give to his permanent records. "There was so much to write," Harry had said.[56]

NOTES

This chapter is adapted from *Hemingway's Art of Nonfiction* (New York: St. Martin's, 1990; London: Macmillan, 1990)

1. Malcolm Cowley, "A Farewell to Spain," *New Republic* 73:76 (November 30, 1932).

2. Ernest Hemingway, note to "A Natural History of the Dead," in *Winner Take Nothing* (New York: Scribner's, 1933), p. 137.

3. Ernest Hemingway, *Death in the Afternoon* (New York: Scribner's, 1932), p. 2.

4. Ernest Hemingway, "A Situation Report," in *By-Line: Ernest Hemingway,* William White, ed. (New York: Scribner's, 1967), p. 472. Ernest Hemingway, "Monologue to the Maestro: A High Seas Letter," in *By-Line: Ernest Hemingway,* p. 215.

5. Ernest Hemingway to Maxwell Perkins, December 6, 1926, in *Ernest Hemingway: Selected Letters, 1917–1961.* Carlos Baker, ed. (New York: Scribner's, 1981), p. 236.

6. Ernest Hemingway, *The Dangerous Summer* (New York: Scribner's, 1985), p. 82.

7. Ernest Hemingway, "On Writing," in *The Nick Adams Stories* (New York: Scribner's, 1972), p. 246.

8. George Plimpton, "Ernest Hemingway," in *Writers at Work: The Paris Review Interviews,* Second Series (New York: Viking, 1963), p. 233.

9. Ernest Hemingway to Maxwell Perkins, April 15, 1925, *Selected Letters,* p. 156.

10. Maxwell Perkins to Ernest Hemingway, November 23, 1926, Scribner Archives, Firestone Library, Princeton University.

11. Ernest Hemingway to Maxwell Perkins, December 15, 1929, *Selected Letters,* p. 317.

12. Ernest Hemingway to Maxwell Perkins, June 2, 1932, Hemingway Collection, John Fitzgerald Kennedy Library, Boston.

13. Edmund Wilson, "Hemingway: The Gauge of Morale," in *The Wound and the Bow* (Cambridge: Houghton Mifflin, 1941), pp. 223, 227. Wilson first broached this view in a discussion of *Green Hills of Africa,* "Letter to the Russians about Hemingway," *New Republic* 81:137–38. (December 11, 1935), and subsequently in "Ernest Hemingway: Bourdon Gauge of Morale," *Atlantic Monthly,* 164:36–46 (July, 1939). Wilson's view of Hemingway's nonfiction was elaborated in a 1939 *Partisan Review* article by Lionel Trilling, who held that while Hemingway the "artist" is conscious, innocent, disinterested, and truthful, Hemingway the "man" is selfconscious, naïve, has a personal axe to grind, and falsifies. The nonfiction, stemming from the "man," would seem to be left with little to recommend it. Trilling, "Hemingway and His Critics," in *Ernest Hemingway.* Harold Bloom, ed. (New York: Chelsea House, 1985), p. 7.

14. Wilson, *The Wound and the Bow,* p. 223.

15. Ernest Hemingway, *The Sun Also Rises* (New York: Scribner's, 1926), p. 245.

16. Lincoln Kirstein, "The Canon of Death," reprinted in *Ernest Hemingway: The Man and His Work,* John K. M. McCaffery, ed. (Cleveland: World, 1950), p. 60.

17. Ernest Hemingway to Maxwell Perkins, December 6, 1926, *Selected Letters,* p. 236.

18. Hemingway, *Death in the Afternoon,* p. 487. Subsequent page references appear in the text.

19. Ernest Hemingway to Maxwell Perkins, December 6, 1926, *Selected Letters*, pp. 236–37.

20. F. O. Matthiessen, *American Renaissance* (New York: Oxford University Press, 1941), p. 170.

21. Thoreau is quoted in Matthiessen, *American Renaissance*, p. 85. The passage appears in *A Week on the Concord and Merrimack Rivers* (New York: The Library of America, 1985), p. 226.

22. Cyril Connolly, *Enemies of Promise* (London: Andre Deutsch, 1973), p. 66.

23. Ernest Hemingway, *Green Hills of Africa* (New York: Scribner's, 1935), pp. 108–9. Subsequent page references appear in the text.

24. Ernest Hemingway to Maxwell Perkins, November 20, 1935, Hemingway Collection, John Fitzgerald Kennedy Library.

25. William Kennedy, "The Last Olé," *New York Times Book Review* (June 9, 1985), p. 32.

26. Ernest Hemingway to Ivan Kashkin, January 12, 1936, *Selected Letters*, p. 431.

27. Ernest Hemingway, "The Snows of Kilimanjaro," in *The Short Stories of Ernest Hemingway* (New York: Scribner's, 1953), p. 59.

28. Philip Young and Charles W. Mann, *The Hemingway Manuscripts: An Inventory* (University Park: The Pennsylvania State University Press, 1969), p. 6.

29. Mary Hemingway, *How It Was* (New York: Knopf, 1976), pp. 532, 535.

30. Quoted by Ray Cave, "Introduction to An African Journal," *Sports Illustrated*, 35:41 (December 20, 1971).

31. Carlos Baker, *Ernest Hemingway, A Life Story* (New York: Scribner's, 1969), p. 526.

32. The *Sports Illustrated* installments appeared on December 20, 1971, and January 3 and 10, 1972. The first two installments were subtitled "Miss Mary's Lion"; the third "Imperiled Flanks."

33. Ernest Hemingway, "African Journal," *Sports Illustrated* 35:13 (December 20, 1971).

34. José Luis Castillo-Puche, *Hemingway in Spain* (Garden City, NY: Doubleday, 1974), pp. 318–19.

35. Ernest Hemingway to Gianfranco Ivancich, May 30, 1960, *Selected Letters*, p. 903.

36. Ernest Hemingway to Bill Lang, January 3, 1960, Hemingway Collection, John Fitzgerald Kennedy Library. *Life* quoted from the letter in an editorial note accompanying the first installment of "The Dangerous Summer."

37. In his introduction to *The Dangerous Summer* James Michener says the manuscript originally had 120,000 words and the edited version submitted to *Life* about 70,000.

38. The holograph manuscript and the typescript of *The Dangerous Summer* are among the Hemingway papers at the John Fitzgerald Kennedy Library.

39. Hemingway, *The Dangerous Summer*, p. 82. Subsequent page references appear in the text.

40. James Michener, Introduction to *The Dangerous Summer*, p. 16.

41. See, for example, Gerry Brenner, "Are We Going to Hemingway's *Feast?*",

American Literature, 54:529 (December 1982), and Jacqueline Tavernier-Courbin, "The Manuscripts of *A Moveable Feast,*" in *Hemingway Notes* [renamed *The Hemingway Review*], 1:9 (Spring, 1981).

42. Ernest Hemingway, "Pamplona Letter," *The Transatlantic Review* 2:301 (September, 1924).

43. Ernest Hemingway to Janet Flanner, April 8, 1933, *Selected Letters,* p. 388.

44. Ernest Hemingway to Maxwell Perkins, July 26, 1933, *Selected Letters,* p. 396.

45. Ernest Hemingway, "The Snows of Kilimanjaro," pp. 70, 71.

46. Ernest Hemingway, "African Journal," *Sports Illustrated* 36:3 (January 3, 1972).

47. For doubt about the existence of the trunks see Jacqueline Tavernier-Courbin, "The Mystery of the Ritz Hotel Papers," in *Ernest Hemingway, The Papers of a Writer,* Bernard Oldsey, ed. (New York: Garland, 1981), pp. 117–31.

48. Mary Hemingway, "The Making of a Book: A Chronicle and a Memoir," *New York Times Book Review* (May 10, 1964), p. 27. Carlos Baker finds Mary's recollection questionable. He points out that Hemingway was not writing a biography, and that the term *remate* in jai alai refers to a kill-shot, one that cannot be played by an opponent. Baker, *Hemingway, the Writer as Artist,* fourth ed. (Princeton: Princeton University Press, 1972), pp. 375–76*n*. Gerry Brenner, however, suggests that Mary had the term right, Hemingway intending his acid portraits of contemporaries in *A Moveable Feast* as "kills-shots they are literally unable to return." Brenner, *Concealments in Hemingway's Works* (Columbus: Ohio State University Press, 1983), p. 218.

49. Ernest Hemingway to General Charles L. Lanham, September 18, 1958, Hemingway Collection, John Fitzgerald Kennedy Library.

50. A. E. Hotchner, *Papa Hemingway* (New York: Random House, 1966), pp. 285–86.

51. Ernest Hemingway to Charles Fenton, July 29, 1952, *Selected Letters,* p. 776.

52. Ernest Hemingway, *A Moveable Feast* (New York: Scribner's, 1964), p. 30. Subsequent page references appear in the text.

53. Marvin Mudrick [review of *A Moveable Feast*], reprinted in *Hemingway: The Critical Heritage,* Jeffrey Meyers, ed. (London: Routledge & Kegan Paul, 1982), p. 509.

54. Mary Hemingway, "The Making of a Book: A Chronicle and a Memoir," p. 27.

55. Maxwell Perkins to Ernest Hemingway, November 3, 1932, Scribner Archives, Firestone Library, Princeton University.

56. Hemingway, "The Snows of Kilimanjaro," p. 66.

3

The Mother of Literature: Journalism and *The Grapes of Wrath*

William Howarth

> What can I say about journalism? It has the greatest virtue and the greatest evil. It is the first thing the dictator controls. It is the mother of literature and the perpetrator of crap. In many cases it is the only history we have and yet it is the tool of the worst men. But over a long period of time and because it is the product of so many men, it is perhaps the purest thing we have. Honesty has a way of creeping in even when it was not intended.
>
> John Steinbeck, letter to the U.S. Information Service
> (L256)

THE RISING WATER

At the end of *The Grapes of Wrath,* natural and human events impel the novel to a relentless climax. Far out at sea early winter storms rise, sweep landward, and pour drenching rain on the California mountains. Streams cascade down into river valleys, flooding the lowlands where thousands of migrant families have set up makeshift camps. Many flee, others resist—and lose their meager goods to the rising water. Having no work or wages till spring, the migrants face a hopeless situation. They begin to starve, dying from exposure and disease, but no relief arrives. The Joad family faces an added crisis, as their daughter, Rose of Sharon, suffers through hard labor and delivers a stillborn child. Water forces the clan to higher ground, where they find a boy and his starving father. The young mother lies down beside the exhausted man. She bares her breast and he feeds.

Although perplexing to generations of readers, that final tableau fulfills a design that governs Steinbeck's entire novel.[1] His book opens with drought and ends with flood, waters that return to the earth and replenish its life. In saving a stranger, Rose of Sharon rises from brute survival instinct into a nurturing state of grace: "She looked up and across the barn, and her lips came together and smiled mysteriously" (619). His editor thought this ending was too enigmatic, but Steinbeck replied: "I've tried to make the reader participate in the actuality, what he takes from it will be scaled entirely on his own depth or hollowness. There are five layers in this book, a reader will find as many as he can and he won't find more than he has in himself" (L178–79).

In time readers have found plenty in *The Grapes of Wrath*, calling it a pack of lies, an American epic, and an act of art wrapped in propaganda. This multeity of response bathed the author in ironies. His book denounced capitalism but rang up towering sales; its fame brought him wealth and power yet ruined him for greater work. "There is a failure here that topples all our success" (477), he wrote of the Depression, with words that could also eulogize his own career. Why did this comet rise in his thirty-seventh summer, and what were its literary origins? Critics have tended to cite the modernist traditions of realism or symbolism, either aligning Steinbeck with social ideology—Farrell, Herbst, Wright—or the cultural aesthetics of Dos Passos and Faulkner.[2] Both traditions regard creativity as a lonely, heroic struggle, fought by artists for the sake of their race. Robert DeMott speaks for this consensus in calling *The Grapes of Wrath* "a private tragedy," in which the writer sacrificed "the unique qualities . . . that made his art exemplary" (Wxlvi) to create a broad social novel.[3]

That vision of martyred demise may explain Steinbeck's late career, but it refutes the meaning of his greatest triumph. The novel's final scene is not a rite of sacrifice but fulfillment, as individual striving gives way to shared alliance. Two persons become one, not through sex or even love, but through their selfless flow into a broader stream, the rising water of human endurance. The novel repeats this idea in many contexts, most notably through Jim Casey, a prophet who fuses the socialist vision of class struggle with a sacramental longing for universal communion: "But when they're all workin' together, not one fella for another fella, but one fella kind of harnessed to the whole shebang—that's right, that's holy" (110). If these ideas hark back to Depression-era politics, they also anticipate a world that has yet to come, tied in bonds of ecological affinity.

While Steinbeck praised collective action he was no doctrinaire socialist, and critics have therefore placed him in a liberal ideology that regards the artist as both seer and entrepreneur: "The communal vision of *The Grapes of Wrath* began in the sweat of Steinbeck's lonely labor" (Wxxxii), writes Demott, whose edition of the novel's "work diary" and Jackson Benson's definitive biography now give us a comprehensive view of the writer and his milieu. Their scholarship also reveals that Steinbeck's labors were far from lonely. In addition to literature he drew constantly from the well of journalism, which he once called "the mother of literature" (L256). This debt was evident to Joseph Henry Jackson, who in 1940 first noted how *The Grapes of Wrath* borrowed its techniques from newsreel, photo-text, radio drama, and proletarian fiction—the peculiar hybrid forms of art, journalism, and propaganda that James Boylan calls "Depression reportage." Reactions to that heritage fueled early quarrels over the book's literal accuracy and the New Critical emphasis on its mythology and philosophy. Another view is now in order, one suggesting that the mother of this novel—and its last maternal scene—was something Steinbeck knew as documentary.[4]

Documentary is a term used since the 1920s to denote the wedding of reportage, the investigative methods of journalism and sociology, to new forms of mass-media imagery, especially photography. Documentary is a didactic art that aims to look hard but feel soft, to affect an audience's emotions with ocular proof, the arrangement of apparently unselected scenes. The style tends to flourish in periods of grave social crisis, traumas that fracture public trust and arouse a clamor for indisputable facts. In this century the impulse has come in twenty- to thirty-year waves, from pre-World War I Muckraking to the New Journalism of the Vietnam era. The Depression stands as a clear highwater mark, for during the 1930s there arose, in the words of historian William Stott, "a documentary imagination," determined to spread truth, right wrongs, and shape a new social order—one governed by the values of equity and cooperation. When Steinbeck alluded to the "five layers" in his book he was professing a documentary faith in data (he later said *Sea of Cortez* had four layers, L232), and in the emotional power of visual effects. Most important of these were his strong images, which emulated not the traditional arts of painting or sculpture but two popular forms of mass media, still and motion photography.[5]

The final chapters of *The Grapes of Wrath* exemplify this method by working through an intricate course of narrative "shots," from opening wide-angle panoramas of sea, air, and land to a tracking montage that follows the floodwaters' descent. This sequence of long- and mid-range images gathers into a fluent cascade of words:

> The rain beat on steadily. And the streams and the little rivers edged up to the bank sides and worked at willows and tree roots, bent the willows deep in the current, cut out the roots of cottonwoods and brought down the trees. The muddy water whirled along the bank sides and crept up the banks until at last it spilled over, into the fields, into the orchards, into the cotton patches where the black stems stood. Level fields became lakes, broad and gray, and the rain whipped up the surfaces. Then the water poured over the highways, and cars moved slowly, cutting the water ahead, and leaving a boiling muddy wake behind. The earth whispered under the beat of the rain, and the streams thundered under the churning freshets. (589–90).

At the river bottoms, these visual effects swiftly change from motion to contrast, an effect achieved by cross-cutting between the flood and the Joad family. Their faces appear in tight shots—portraits and extreme closeups—that sharply differ from the epic anonymity of landscape and "the people." Such juxtapositions suggest equivalence, a strong parallelism between large and small events, the seen and unseen. As the Joad women struggle to deliver a child, their men labor to erect an earthen dike around a boxcar shelter. They succeed until a great cottonwood tree floats downstream, snags, and swings toward them:

> The water piled up behind. The tree moved and tore the bank. A little stream slipped through. Pa threw himself forward and jammed mud in the break. The water piled against the tree. And then the

bank washed quickly down, washed around ankles, around knees. Then men broke and ran, and the current worked smoothly into the flat, under the cars, under the automobiles.

Uncle John saw the water break through. In the murk he could see it. Uncontrollably his weight pulled him down. He went to his knees, and the tugging water swirled about his chest. . . .

When the dike swept out, Al turned and ran. His feet moved heavily. The water was about his calves when he reached the truck, He flung the tarpaulin off the nose and jumped into the car. He stepped on the starter. The engine turned over and over, and there was no bark of the motor. He choked the engine deeply. The battery turned the sodden motor more and more slowly, and there was no cough. Over and over, slower and slower. (601–2)

The visible events—snagged tree, breached dike, dead engine—intimate an unseen drama transpiring in the boxcar, as Rose of Sharon labors through her stillbirth. Narration is thus working in two focal planes, exterior foreshadowing interior, with Steinbeck shifting from natural panorama to human close-up, one cause to another effect:

The air was fetid and close with the smell of the birth. Uncle John clambered in and held himself upright against the side of the car. Mrs. Wainwright left her work and came to Pa. She pulled him by the elbow toward the corner of the car. She picked up a lantern and held it over an apple box in the corner. On a newspaper lay a blue shriveled little mummy.

"Never breathed," said Mrs. Wainwright softly. "Never was alive."

Uncle John turned and shuffled tiredly down the car to the dark end. The rain whished softly on the roof now, so softly that they could hear Uncle John's tired sniffling from the dark. (603)

Rainfall merges with tears, just as the floodwaters will give way to a mother's milk. Throughout the novel Steinbeck oscillates from large- to small-scale events, building a dialectic between nature and humanity, the masses and the Joad family. This synecdochal linkage is an ancient narrative device, common to folk and scriptural sagas, that gives *The Grapes of Wrath* its epic dimensions. The principle is also visual, moving from distance to foreground and back again to provide pictorial scale and depth. Steinbeck was quite aware of this formal method, alluding frequently in his work diary to a plan of alternating "general" and "particular" chapters (W23, 39).[6] That plan arose from a simple paradox: his book really began with its ending.

ON ASSIGNMENT

The final chapters of *The Grapes of Wrath* represent actual events that Steinbeck witnessed in February, 1938, near Visalia, California. Unusually prolonged rain that winter produced high floods along the local streams, stranding more than 5,000 migrant families in homeless destitution. For several days Steinbeck joined relief efforts there, moving families and caring for the sick. People were dying from cholera or starvation, often simply dropping dead in their tracks. He later said the experience hurt inside, "clear to the back of my head" (Wxliii), and the probable source of pain was guilt. After passing some years as a raffish bohemian he had attained the status of best-selling author, enriched by stage and film adaptations of his books. His sympathy for victims of the Depression was clear, but so, too, was the fact that he earned a living as chronicler of their pain: "Funny how mean and little books become in the face of such tragedies" (L159).[7]

In fact, his purpose at Visalia was twofold: to garner publicity for federal efforts at flood relief, and to gather material for a text-and-picture book about California's farm migrants. The latter project was proposed to him in late 1937 by Horace Bristol, a San Francisco photographer with many story credits at *Life*. For several weekends they had traveled throughout the Central Valley, operating as a team to capture interviews and images. Steinbeck was a natural collaborator, Bristol recalled: "He'd get to talking to somebody, and I'd move in behind him with the camera." Using a twin-lens reflex in natural light, Bristol took unposed, candid shots of people at close or mid-range—wading in knee-deep water, standing before tent homes and mired autos. (Figure 3.1) He also took an especially memorable picture of a young mother nursing her baby, an image later widely assumed to be the origin of Rose of Sharon. In all he shot more than 2,000 frames, and in the process he strongly affected Steinbeck's vision of the migrants. In early

Figure 3.1. Horace Bristol. Flooded boxcar home near Visalia, 1938. (Courtesy of Horace Bristol.)

March the writer told his literary agent "we took a lot of pictures," adding that the images "give you an idea of the kind of people they are and the kind of faces" (L161).[8]

After the Visalia experience, Steinbeck brooded on how best to write about those people. Bristol envisioned a book like *You Have Seen Their Faces* (1938), an album on Southern sharecroppers by Erskine Caldwell and Margaret Bourke-White. To publicize this goal, Bristol negotiated with editors at Time, Inc., for a major photo-essay. Steinbeck rejected a lucrative deal with *Fortune* as inappropriate, but he agreed to an offer from *Life* that covered expenses and a donation to flood relief (L162). The team submitted a story, which *Life* chose not to publish at that time. Defiantly, Steinbeck placed a strong news report, called "Starvation Under the Orange Trees," in the *Monterey Trader* (April 15, 1938). By late spring Bristol had selected his pictures for the book, only to find that Steinbeck had other plans: "Well, Horace, I'm sorry to tell you, but I've decided it's too big a story to be just a photographic book. I'm going to write it as a novel."[9]

In a 1951 reminiscence Steinbeck wrote of his father: "To be anything pure requires an arrogance he did not have, and a selfishness he could not bring himself to assume. He was a man intensely disappointed in himself. And I think he liked the complete ruthlessness of my design to be a writer in spite of mother and hell" (W140). Back in mid-1938 Horace Bristol had seen a flash of the writer's arrogance and its ruthless designs. From then on Steinbeck rarely acknowledged working with Bristol or the effect of seeing his migrant pictures. Although Bristol lost or destroyed most of the photographs, he published a surviving handful in 1989. By then he was inclined to forgive, if not forget, Steinbeck's slight: "He was secretive, and he wanted to feel that anything he did was his, and nobody else had any influence on it."[10]

In spurning Bristol, Steinbeck also concealed the fact that since 1936 he had pursued "The Matter of Migrants," as he called plans for a book of his own. He kept this work secret mainly because it was controversial and dangerous. Labor strife pitted the government against California's Associated Farmers, who often hired vigilantes to break strikes and beat investigators. Fearful of betrayal or retaliation, Steinbeck worked undercover and incognito. Using journalistic assignments to underwrite his field trips, he took data wherever he found it—readings, interviews, background briefings—and borrowed stories freely, in the name of public interest. Given these methods, he had trouble deciding whether his book should be factual or fictional, and what his stake was in its creation.[11] Sometimes the goal was selfless advocacy, rather than popularity: "I want this book to be itself with no history and no writer" (L181). At other times ambition led him on, toward the alluring goal of fame. An echo of that conflict resounds in his novel: before leaving Oklahoma, Ma Joad burns her precious box of "letters, clippings, photographs" and keeps a few bits of gold jewelry (148).

THE WRITER AS REPORTER

This wayward journey began when Steinbeck published "Dubious Battle in California," a brief essay about labor migrants in *The Nation* (September 13, 1936), and then agreed to write a series of investigative articles for the San Francisco *News*. He obtained major assistance from Farm Security Administration documents, including reports compiled by Tom Collins, manager of a federal camp for migrants at Arvin. Collins, a psychologist with a flair for writing, had extensively interviewed his residents and gathered a thick compendium of their stories, songs, and folklore, interpolated with his own descriptions and impressions. He generously lent this material to Steinbeck, who used it and his own notes (gathered on trips with Collins) to write "The Harvest Gypsies," a seven-part series for the *News* (Oct. 5–11, 1936). The project won widespread attention, in part for Steinbeck's impassioned, detailed reporting and in part for several pictures shot by Dorothea Lange, an acclaimed FSA photographer.[12]

Steinbeck's association with the FSA gave him contact with a federal agency that effectively used documentary photos to promote New Deal programs. Guided by Roy Stryker, the Historical Section of the FSA compiled more than 270,000 images of American life in 1935–1943, much of it shot by such leading photographers as Lange, Walker Evans, Russell Lee, Arthur Rothstein, Gordon Parks, and Ben Shahn. Stryker set high technical standards for their pictures, calling for an emphasis on individual portraits rather than mobs of striking field hands. The results were often idealized pictures of decent, dignified folk who were blamelessly down on their luck. Dorothea Lange specialized in what William Stott aptly calls "the worthy poor," people of low status whose portraits maintained a quality of intimacy and poignancy for their middle-class viewers. Her style of work embraced the inherent tenets of documentary, authenticity heightened by emotional appeal. With resolute control she chose her subjects, posed them carefully, and supervised the processing of negatives and prints. At the same time she was candid about this work and its purposes, unlike the shy, secretive Steinbeck.[13]

The migrant stories won wide attention in the *News* and even more when reprinted as a pamphlet, *Their Blood is Strong* (1938), by the pro-labor Simon J. Lubin Society. On the cover was Lange's photo of a young woman in a tent shelter, nursing her child (Figure 3.2). The final scene of *The Grapes of Wrath* may have borrowed from this image, which projects a striking aura of ambiguity. Seated in a makeshift cluttered space, the woman looks straight ahead with a calm, impassive expression. Her face and body are youthful, having the strong planes and curves that a lens favors, but her bared breast leads the eye to the central figure, a child's wide-eyed, demanding face. Other shots of this scene include a man in the foreground; in this picture, his legs are partially visible. Linked to Steinbeck's title phrase, the cover shot was designed to stir feelings of racial pride, for his text made clear that the migrants' strong blood flowed in Anglo-Saxon veins. Yet the image also illustrates the ambivalence of documentary: years later, William Stott saw a woman with "the innocent trapped eyes of a deer" while feminist Wendy Kozol saw one "assertive in her direct gaze." Both viewers recognized one of the FSA's stereotypical genres, the nurturing Madonna who appealed strongly to middle-class readers.[14]

Throughout 1937 Steinbeck cast about for his book on the migrants, wavering between the poles of fact and fiction. Impressed with the "great gobs of information" (Wlii) in Tom Collins's camp reports, he offered to help edit them into a diarylike narrative that "would make one of the greatest and most authentic and hopeful human documents I know" (B348). At the same time, Steinbeck envisioned writing an epic novel, tentatively called *The Okla-*

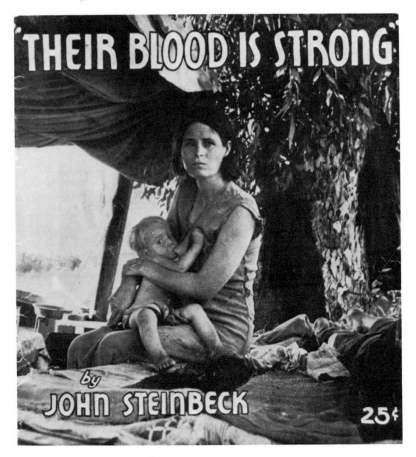

Figure 3.2. Dorothea Lange. Partial front cover of *Their Blood is Strong.* (Courtesy of Steinbeck Research Center.)

homans, about the migrants' journey to California (Wxxxvii). To advance both plans, he and Collins made extensive Central Valley trips in the summer and fall of 1937, visiting squatters' camps and working as field hands. These experiences confirmed the accuracy of Collins's reports, which Steinbeck saw as a source to keep the novel honest, safe from charges of lying (Wxxviii). By winter,

however, both projects subsided and Steinbeck instead accepted
Horace Bristol's proposal to collaborate on a picture book. They
began to travel on weekends, and were sometimes accompanied by
Collins.

When the floods hit Visalia in February, 1938, Steinbeck again
found himself caught between literary and journalistic impulses.
Frustrated by poor coverage in local newspapers, officials at the
FSA asked him to write some reports for national circulation. He
saw this task as a humanitarian mission: "I'm going to try to break
the story hard enough so that food and drugs can get moving. . . .
If I can sell the articles I'll use the proceeds for serum and such"
(L159). During this time negotiations went forward with *Fortune*
and *Life,* and by March he had filed the story. Yet by choosing *Life*
to publicize the migrants, Steinbeck had thrown in his lot with a
photographer's magazine, not a writer's. The pages of *Life* featured huge images, printed with such clarity and depth of focus
that they created the illusion of complete authenticity. The result
was strong propaganda, echoing formulas followed by *Der Spiegel*
and *Stern* in Nazi Germany.

No less strident was Steinbeck's concurrent effort to write a
novel, called *L'Affaire Lettuceberg,* about a farm labor strike. Intended as satire, the project soon degenerated into a bitter tract,
its sarcasm sharply contrasting with the generous spirit of his Visalia story. By May he completed a draft but then declined to
publish it, telling his agent that the work was too biased and incomplete: "Oh! these incidents all happened but—I'm not telling as
much of the truth about them as I know. . . . My whole work drive
has been aimed at making people understand each other and then I
deliberately write this book the aim of which is to cause hatred
through partial understanding. . . . A book must be a life that lives
all of itself and this one doesn't do that" (Wxl).

A book "that lives all of itself" would merge advocacy with
altruism, and this documentary standard had arisen since Steinbeck's experiences in Visalia. The flood scenes and pictures began
to focus his previous attempts to write about migrants. To date, the
Collins reports and his own drafts had yielded three different stories: life in the camps, the long journey west, and strikes. In the
Visalia experience Steinbeck now saw a climactic episode that
fused those materials. A five-part narrative sequence—drought,

flight, camp, strike, and flood—would encompass the migrants' entire history and geography. His story plan thus emerged: alternating chapters, strong visual images, themes of sharing versus greed. Steinbeck's two-year search was over, and in that moment of triumph he cast aside his collaborators.

Tom Collins proved least disposable, for his camp reports were so invaluable that Steinbeck dedicated the novel "To TOM who lived it" and used him as a model for Jim Rawley, manager of the Weedpatch camp. Horace Bristol had to go, however, not because his photographs were inappropriate, but because they were too influential. The claim that Visalia was "too big a story to be just a photographic book" belied the fact that pictures helped Steinbeck find a narrative plan and stylistic model. In addition, if the story was too big for photos, why did it have to be fiction? He had aborted four previous efforts, two novels and two documentaries, in which the elements of actuality and imagination would not coalesce. The migrant story begged for documentation, since many facts were in dispute, but most critics believe that Steinbeck somehow lost his reportorial appetite at Visalia. Joseph Millichap writes, "The reality he encountered seemed too significant for nonfiction," an opinion echoed by Robert DeMott: ". . . he was utterly transfixed by the 'staggering' conditions, and by 'suffering' so great that objective reporting would only falsify the moment" (Wxlii).[15]

Steinbeck refutes that notion of reportage in the very letter cited, attacking "slick" magazines like *Fortune* and *Life,* but not all periodical journalism: "I want to put a tag of shame on the greedy bastards who are responsible for [neglecting the migrants] but I can do it best through newspapers" (L162). Even so, his energy soon shifted to creating the Joad family, a fictional cast that De-Mott says "elevated the entire history of the migrant struggle into the ceremonial realm of art" (Wxliii). This sacramental view corresponds to a general belief that the Joads are mythical questers, but considerable evidence also points to their origin in documentary sources. From Collins, Lange, and Bristol, Steinbeck gleaned several text-and-picture portraits that suggested his family characters. *Life* later confirmed that parallel, by printing Bristol's pictures with captions from *The Grapes of Wrath.*[16]

The notion that Steinbeck turned to fiction because it was a higher, truer from of expression either betrays his intellect or his documentary principles. He chose fiction to make his story more artful, not truthful. In fiction he could fabricate at will, making up people and events by splicing and reshaping materials gathered from research. The decision to anchor his "particular" story in a family of characters was inspired but hardly original, for many documentaries did the same in order to dramatize social history. The Joads were both individual and universal; they gave the story credibility rather than authenticity. As Mark Twain once remarked, "Truth is stranger than Fiction . . . because Fiction is obliged to stick to possibilities. Truth isn't." Yet *The Grapes of Wrath* is often truthful because it strives to emulate documentary genres: case study, informant narrative, travel report, and phototext. Steinbeck wanted his migrant book to be honest and moral, an act of social expiation. As he wrote his agent, "I'm trying to write history while it is happening and I don't want to be wrong" (L162). Instead the book reaped great profit and notoreity, both further obscuring its documentary origins. That source was clearer in June, 1938, as he began the manuscript headed "New Start/Big Writing" and also a daily log.[17]

DIARY AND NOVEL

Of all his working papers—news clips, government briefs, reports, and letters—none were more valuable to Steinbeck than the private journal, or "daily work diary" (W39), that he kept during the five-month process of writing *The Grapes of Wrath*. The diary served several functions, practical and psychological, but it was always more than just a craftsman's tool. Steinbeck attached a totemic significance to his writing habits, from a pen and ruled notebook to his small room and enforced hours of quiet labor. The diary was his gateway to and from this work, allowing him to pass into the narrative and back out to daily life. Writing an entry each morning replaced his usual letters but maintained their aspect of dialogue: "Must make note of work progress at the end of this day. I want to finish my stint if I possibly can. Impulses to do other things. Wind blowing over me, etc." (W21).

He was talking to and of himself, in that subject–object blurring that invokes character, and some of his mood probably emanated from a scene just completed in Chapter 4, where Tom Joad and Jim Casey discuss the conflict between duty and impulse, body and spirit: "Maybe all men got one big soul ever'body's a part of," Casey concludes (33). That was true in the unitary realm of imagination, but in the diary Steinbeck also recorded his schisms, moments when all the world's distractions—friends, talk, drink, sex, money, news—sundered his concentration, broke his confidence in the story. The diary offered a disciplinary ritual, each day launching, sustaining, and focusing his efforts, admitting him to a realm where reality and fiction were coeval: "In ten days I will be half done. 50 days of work. I hope we get to California by then because I would like half the time out here" (W46). The "I" and "we" referred to himself and his characters, or alternately, to writer and diary. In the isolation of his work, its pages became his final collaborator.

Besides documenting his progress the diary shaped its direction, turning his writing into a journey that paralleled the Joads'. Much of the plot outline arose from his decision to begin in Oklahoma, for that impelled the story to move constantly forward, each chapter rolling west along "the main migrant road" to reach California (160). The "unity feeling" (W27) Steinbeck often sensed while writing arose from identifying his own motion, line by line and page by page, with the miles slipping past his characters. He calculated time and distance in words, not cash or fuel, striving "to try to maintain a certain writing speed" (W25). At times writing was like Tom Joad's prison term, "Must not think too much of the end but of the immediate story—instant and immediate" (W38), for in each given moment of composition, the story had a fluid momentum of its own: "And so the book moves on steadily, forcefully, slowly, and it must continue to move slowly. How I love it" (W31).

This sense of inevitability encouraged him to emulate the characters at many levels, sharing their losses, determined to survive by going on: "Can't tell. Can't tell. Just have to plod for 90,000 words. Plod as the people are plodding. They aren't rushing" (W59). His dogged fatalism advanced him in a fairly straight line, with fewer stalls and breakdowns than the migrants, though at times he puzzled over how plot sequences "mapped out" (W72) or he turned from linear progress to make adjustments: "Must go back and send Noah down the river" (W73). And so the writing went, fast or slow, mirroring the Joads' progress and enhancing their story with the mythic aura of westering, the American journey sustained by an indomitable work ethic: "I'll get the book done if I just set one day's work in front of the last day's work. That's the way it comes out. And that's the only way it does" (W85). He was then writing Chapter 26, where the martyred Jim Casey asserts that human progress is inevitable (525), echoing Ma Joad's earlier sentiment, "Why, we're the people—we go on" (383).

The diary was not fiction, but a fact-book; it monitored his progress, corrected errors, curbed the impulse to stray from sources. He sensed quite early that this arbiter would hold him to documentary standards: "I have tried to keep diaries before but they don't work out because of the necessity to be honest. In matters where there is no definite truth, I gravitate toward the opposite" (W19). This diary linked those opposites, allowing him to enter and exit the story, to write alternating units—"Yesterday the general and now back to the particular" (W23)—and to rein in his imagination: "The more I think of it, the better I like this work diary idea. Always I've set things down to loosen up a creaking mind but never have I done so consecutively. This sort of keeps it all corralled in one place" (W42). Discipline yielded force, but not always conscious control. "I did it but it may not be good. I don't know. But it is in, the I to We" (W43). He had just finished Chapter 14, on the growth of revolutionary masses, and the same transfer of power was beginning to affect his book: "For the quality of owning freezes you forever into 'I,' and cuts you off forever from the 'we' " (206).

If Steinbeck did not live by that ideal (he was then buying a new ranch, the biggest property deal of his life—W15),[18] he at least wrote toward it, for the diary and book had acquired an autonomy, in which he participated as both inventor and recorder: "Often in writing these beginning lines I think it is going to be all right and then it isn't. Just have to see. I hope it is all right today" (W46). Although his life was especially fractious at this time, his work had an energetic spontaneity of its own: "Where has my discipline gone? Have I lost control? Quite coldly we'll see today. See whether life comes into the lives and the people move and talk. We'll see. Got her, by God" (W61). Somewhere between seeing and shaping lay the story. He could push characters about and make events happen, but also just open himself to intuitive possibilities: "Got to get them out of Hooverville and into a federal camp for they must learn something of democratic procedure. . . . The flow of story is coming back to me. The feel of the people. And the feel of speech and the flow of action. So it must go and if it takes until Christmas, then that is the way it must be" (W64–65).

As the story reached its late stages, narrative momentum became a compelling force. His daily exhortations took on a quality of passionate testimony: "This is the important part of the book. Must get it down. This little strike. Must win it. Must be full of movement, and it must have the fierceness of the strike. And it must be won" (W79). Of course the story was not writing itself, but he cast himself increasingly as its witness or reporter, just giving an account of what passed before his eyes: "And my story is coming better. I see it better. See it better. . . . Half an hour gone already and I don't care because the little details are coming, are getting clearer all the time" (W82–83).

PICTURES STILL AND MOVING

The work diary often reveals how persistently Steinbeck described his imagination in photographic terms, as though his mental images arose from an unmediated source. Like pictures in a tray of developer, the final scenes emerged slowly but with startling clarity. He had known from the outset how and where the story would end, but getting there proved difficult, for he had incorrectly calculated the term of Rose of Sharon's pregnancy (W72). If she were to deliver at full term in winter, that would mean conception back in the spring, before the novel begins. He fixed this by having Pa vaguely announce, "An' Rosasharn's due 'bout three-four-five months now" (113) and later describing the stillborn infant as premature, "a blue shriveled little mummy" (603). Nervous doubts arose about the flood and birth sequence (W86), but he resolved them by letting his plan emerge from the characters' needs: "The last general must be a summing of the whole thing. Group survival. Yes, I am excited. Almost prayerful that this book is some good. Maybe it is and maybe not. Now let's see what we have" (W88).

That "we" was a group that survived through labor and love, for in coming to closure Steinbeck felt a mixture of "plain terror" and "some kind of release" at his prospect of loss (W92). To pass this barrier he reread the work diary, seeing in its pages a visual image of his creative process: "And I think I have every single move mapped out for the ending. I only hope it is good. It simply has to be. Well, there it is, all of it in my mind" (W92). This mental habit may reflect the continuing affect of Horace Bristol's pictures, later recognized by *Life* as a principal source for *The Grapes of Wrath:* "Never before had the facts behind a great work of fiction been so carefully researched by the newscamera." Robert DeMott insists that the writer and camera saw events quite differently, right up to the final scene. He argues that Bristol's picture is merely of a mother nursing an infant, while Rose of Sharon gives her breast to a starving man, "a leap beyond facticity" (Wlvi) (Figure 3.3).

That term forces a distinction, since the usual signifiers (*factitious* and *fictitious*) both mean contrived, made artifically. Like all photographs, Bristol's picture is not reality but its representation, flattened to the constraints of two dimensions and one comment. As Susan Sontag notes, photographs are artifacts that also seem "to have the status of found objects—unpremeditated slices of the world." Bristol's image of the nursing mother may appear to be self-evident, yet its formal aspects—angle, elevation, focus, depth, frame, light and shadow—all reflect his work as a selective, shaping recorder.[19] In substituting a man for a baby Steinbeck slipped into double exposure, superimposing memories of other photos (Lange's no doubt included) on scenes he witnessed at Visalia: "[I]n our agricultural valleys, I've seen a family that was hungry give all its food to a family that was starving. I suppose that is inspiration. It and things like it only make me feel like a rat" (W142). Such images of instinctive generosity compelled him to foresee the entire book, including "the last scene, huge and symbolic, toward which the whole story moves," as a humbling revelation: "I felt very small and inadequate and incapable but I grew again to love the story which is so much greater than I am" (W36).

This attitude led him to see the book in iconic terms, consistent with the values of modern photojournalism. The persistent focus on cars and highways matched FSA standards,[20] as did the alternation of near and far views: "This is a huge job. Musn't think of its largeness but only of the little picture while I am working. Leave

Figure 3.3. Horace Bristol. Young mother and newborn, Visalia, 1938.
(Courtesy of Horace Bristol.)

the large picture for planning time" (W29). To him the characters
were figures who moved in and out of focus: "Watch the old people. Might get out of hand but I want them mean and funny"
(W30), and in those two dimensions they shifted constantly between his needs and theirs: "Make the people live. Make them
live. But my people must be more than people. They must be an
over-essence of people" (W39). They should be *and* mean, existing
apart but also within him. Hence their departures aroused his
fierce empathy—the death of Grampa Joad made Steinbeck write
". . . once this book is done I won't care how soon I die" (W41),
while Tom Joad's farewell induced an apparent hallucination: "I
hope the close isn't controlled by my weariness. . . . 'Tom! Tom!
Tom!' I know. It wasn't him. Yes, I think I can go on now. In fact, I
feel stronger. Much stronger. Funny where the energy comes
from" (W91).

The energy of *The Grapes of Wrath* came not just from still
photography but also motion pictures, a source that Steinbeck
more openly acknowledged as influential. Principles of cinematic
narrative sprang directly from fiction, and by 1938 Steinbeck had
absorbed enough movies to recognize their enormous power to
move and inform. Most of his 1930s novels have strong filmic
qualities, which accounts for their swift adaptation into screenplays, work that he assisted by suggesting locations and reading
draft treatments. Film rights to *Of Mice and Men* sold easily, but in
the early stages of writing *The Grapes of Wrath* he waved off an
inquiry from the Selznick company as premature, telling his agent
"no picture company would want this new book whole" because it
was too controversial (L168). In the 1930s Hollywood did offer
him many liberal friends, including the actors Wallace Ford, Brod-
erick Crawford, and Charlie Chaplin, but the most influential film-
maker of all was Pare Lorentz, the brilliant documentarian whom
Steinbeck first met in January, 1938, just prior to going to Visalia.[21]

Lorentz was anything but Hollywood, for he created highly original films (and radio plays) about such unpromising subjects as soil
conservation and flood control. As head of the U.S. Film Service
from 1935 to 1940, he worked on low budgets but managed to
orchestrate the talents of top writers, photographers, and direc-
tors. Two of his films, *The Plow that Broke the Plains* (1936) and
The River (1937) established new standards in American documen-

tary, bringing lyricism and epic scale to the promotion of New Deal programs. Lorentz made his films encompass vast sweeps of history, from the settlement of the prairies to the Depression crisis, and he was able to dramatize vividly how economic and social forces had shaped the American land and people. *The River* was his tour de force, employing imagery and sound in complex patterns to describe the Mississippi Basin and its ecological fate. Especially effective was Lorentz's incantatory text, simple phrasings intoned by Thomas Chalmers, a Met baritone, to accompany a fluid montage of images and music scored by Virgil Thompson.[22]

Lorentz and his films were on Steinbeck's mind throughout early 1938, as he groped with his various writings on migrants. Lorentz helped midwife *The Grapes of Wrath* by placing "Starvation Under the Orange Trees" in the Monterey paper, and then dissuading Steinbeck from further work on his strike novel, *L'Affaire Lettuceberg* (B371–72, W151). The two men also talked of film collaboration, either by adapting a Steinbeck novel such as *In Dubious Battle,* or by producing an original screenplay. Yet as Steinbeck pursued his "Big Writing" he developed characteristic hesitations. In August Lorentz read early chapters and praised them highly, which encouraged the writer to press on—and, as he wrote in the diary, to stay clear of films: "I do hope that Pare won't need me, much as I'd like to work with him. I must do my own work and I have a feeling that he is this picture not me" (W65).

Just as he had withdrawn from Collins and Bristol, Steinbeck evaded Lorentz while absorbing the effect of his style and ideas. One of Lorentz's radio dramas closed with a stirring rendition of "The Battle Hymn of the Republic," which may have suggested *The Grapes of Wrath* as a title to Steinbeck's wife, Carol (Wxxvi). The grateful author later rewarded her with a dedication, "To CAROL who willed this book," for on one level he regarded his novel as autogenous, self-created and sustaining: "The looks of it—that marvelous title. The book has being at last" (W65). His fiction also emulated Lorentz's narrative principles, shifting from foreground to background, cutting from panorama to close-up, and providing choric and lyric commentary in the expository chapters. Steinbeck later acknowledged that his most conspicuous borrowing was for Chapter 12, the naming of towns along Route 66: "I have little doubt that the Lorentz River is strong in that" (Wli).

The work diary reflects other parallels, for he frequently described the novel in cinematic terms, as pictures or scenes that move: ". . . the movement is so fascinating that I don't stay tired. . . . slow but sure, piling detail on detail until a picture and an experience emerge. Until the whole throbbing thing emerges" (W25). He played this action out in settings, the physical locations—porch, road, gas station—that gave strong scenic values to his themes. Chapter 15 begins in what Steinbeck called "an important place" (W44), a hamburger stand that he describes entirely in sentence fragments. Action and continuity in the sequence come from visual tracking rather than verbs:

> At one end of the counter a covered case; candy cough drops, caffeine sulphate called Sleepless, No-Doze; candy, cigarettes, razor blades, aspirin, Bromo-Seltzer, Alka-Seltzer. The walls decorated with posters, bathing girls, blondes with big breasts and slender hips and waxen faces, in white bathing suits, and holding a bottle of Coca-Cola and smiling—see what you get with a Coca-Cola. Long bar, and salts, peppers, mustard pots, and paper napkins. Beer taps behind the counter, and in back the coffee urns, shiny and steaming, with glass gauges showing the coffee level. And pies in wire cages and oranges in pyramids of four. And little piles of Post Toasties, corn flakes, stacked up in designs. (208)

The passage illustrates why Steinbeck chose to write fiction rather than documentary, for despite all its concrete detail this place is imaginary, a broad, generic description that is "typical" instead of literal.[23] Suppression of the verbs also disrupts normal syntactic relationships between subject and object, so that the scene plays not through a narrator or character, but a transparent viewer. The result is prose that emulates film, less from the perspective of a director or editor than from a sensitive observer, which at the time is mainly how Steinbeck knew movies.

In writing *The Grapes of Wrath* he drew closer to film techniques, even while dodging requests to write screenplays for Lorentz or actor Paul Muni, who wanted to adapt *Tortilla Flat* (W76). The late chapters became especially cinematic in style, as they picked up the writer's sense of driving momentum: "Just a little bit every day. A little bit every day. And then it will be through. And the story is coming to me fast now. And it will be fast from now on. Movement fast but the detail slow as always" (W83). While he had brief doubts about including the flood (W86), he resolved them by creating a sequence that imitates the opening and middle parts of *The River,* where water rises high in the mountains and tumbles down to drowned valleys (589ff). Letters from Lorentz arrived as Steinbeck wrote this section, rousing thoughts of a film and doubts about his novel: "The rain—the birth—the flood—and the barn. The starving man and the last scene that has been ready so long. I don't know. I only hope it is some good. I have very grave doubts sometimes. . . . I am sure of one thing—it isn't the great book I had hoped it would be. It's just a run-of-the-mill book. And the awful thing is that it is absolutely the best I can do" (W88–90).

Some of this despairing mood came from exhaustion and a performer's lack of perspective, but Steinbeck also sensed a disparity between his story and talent. In February the floods at Visalia had filled him with anguish; now in mid-October they threatened once more. With so much material at hand, sources compiled and edited into a new order, he carried a lonely burden: "Well, I might as well get to the work. No one is going to do it for me" (W92). Yet in mounting the drive toward completion, he saw himself mainly as a reporter: "Forget that it is the finish and just set down the day by day work. . . . Best way is just to get down to the lines. . . . Finished this day—and I hope to God it's good" (W92–93). This ambivalence about his role may account for a later remark, that he wrote *The Grapes of Wrath* "in a musical technique" governed by mathematical principles, for such abstractions would also limit his authorial responsibility (W13).

AFTERMATH

In the months that followed completion of his "Big Writing," Steinbeck turned with relief back to collective ventures. Filming of *The Grapes of Wrath* in late 1939 involved him in a reprise of the novel's creation: Tom Collins worked as a technical advisor, while director John Ford consulted Lange's photos and Lorentz's films for stylistic pointers. Even Horace Bristol finally got some attention because *Life* published his photos to verify the accuracy of both novel and film. As Joseph Millichap confirms, Ford's film imitates the look of documentary rather than its feel, softening Steinbeck's political themes by emphasizing nostalgia over anger, the Joads over the masses, and eliminating entirely the novel's general chapters.[24] Yet after its opening in January, 1940, it was an enormously popular film and undoubtedly helped to prolong the novel's public recognition.

The firestorm of protest that greeted *The Grapes of Wrath*, most of it raging over the book's "truth," ultimately eroded its efficacy as a social document. Public interest in rural poverty waxed and soon waned, creating little market for such documentary treatments as Carey McWilliams, *Factories in the Field* (1939), Dorothea Lange and Paul Taylor, *An American Exodus* (1939), or James Agee and Walker Evans, *Let Us Now Praise Famous Men* (1940).[25] Yet celebrity gave Steinbeck himself many opportunities to pursue documentary projects. He did finally collaborate with Lorentz on *The Fight for Life* (1939), a screenplay about infant mortality, and then wrote both story and script for *The Forgotten Village* (1941), on the coming of modern medicine to rural Mexico. In his preface to the film's published text Steinbeck justified his use of fictional devices, such as telling the story through one family, asking villagers to re-enact events, and providing choric commentary through an offscreen narrator, "so natural and unobtrusive that an audience would not even be conscious of it."[26]

After that came *Sea of Cortez* (1941), an account of a marine biological expedition that Steinbeck edited from trip logs, adding considerable form and poetry of his own, according to "coauthor" Ed Ricketts (W104–5). World War II then swept in, bringing years when tumult compelled the writer back into a welcome role of social duty, where he wrote training films and dispatches for a cause that had few detractors. His remaining years were to be crowded and productive, often with projects that featured reporting and photography. He never again reached the highwater mark of *The Grapes of Wrath*, a book that began in flood and ended with a vision of the enduring human family.

In moving toward that close, the novel passes through its alternating phases of wind and water, fruition and decay, and junked cars and the relentless "fury of work" (599) that drives people forward, ever determined to survive despite the constancy of gain and loss. Beginning in the center of America, the migrants find at the continent's edge that Eden still lies elsewhere, down the road that pulls them with the relentless tug of a future. While alternating chapters give *The Grapes of Wrath* its sweep and cadence, they also accentuate its major flaw: Steinbeck's failure to integrate fully the modes of journalism and literature.

His "general" chapters are too often simplistic editorials, haranguing the reader with dire prophecy, crude analysis, and crypto-philosophy that the "particular" chapters cannot sustain. In addition, his fictional characters are too often mouthpieces for gradiose ideas, stereotypes that speak in "folk" lingo or stand as rigid symbols of portentous ritual. Yet Steinbeck believed that his documentary sources would give him a great living story, and he was not wrong. Despite its melodrama and bombast, the novel endures by drawing strength from its origins. At one moment, the book even reflects on this process of creation: "And it came about in the camps along the roads, on the ditch banks beside the streams, under the sycamores, that the story teller grew into being, so that the people gathered in the low firelight to hear the gifted ones. And they listened while the tales were told, and their participation made the stories great" (444).

That vision of art in the service of others captures the novel's deepest conviction—that between the general and the particular lie undissolvable bonds. For all its roughhewn quality the Joad family remains a collective unit, able to function best when sharing thought and labor. Their trials on the road and in squatter camps simply enlarge that principle, as adversity binds families with neighbors, turning migrants into a nomadic people: "In the evening a strange thing happened: the twenty families became one family, the children were the children of all. The loss of home became one loss, and the golden time in the West was one dream" (264). The Joad family gradually dwindles in size. Elders die, sons drift away, and infants come and go. Each of these losses impels the survivors toward a larger sense of union, beyond their original clan and tribe. At the end Jim Casey and Tom Joad may be gone, but they have passed a legacy of communion to Ma. After the stillbirth she tells Mrs. Wainwright, "Use' ta be the fambly was fust. It ain't so now. It's anybody. Worse off we get, the more we got to do" (606). That idea passes from mother to daughter, then takes an incarnate form when Rose of Sharon gives her breast to a stranger.

In working toward that final scene, conceived so early in his imagination, Steinbeck himself struggled through a long gestatory process that often threatened to abort or end in stillbirth. Along the way, many of his attitudes toward sexual and familial roles evolved, altering his sense of the project's identity. His early vision of documentary as an act of coalescence, a shared alliance through selfless communion, was naïve and adolescent, attributing greater dialectical and collaborative powers to Lange, Collins, Bristol, or Lorentz than they actually possessed. At the same time Steinbeck denied their spirit of cooperation by insisting on an image of artistry as stereotypically masculine. He believed that his father "liked the complete ruthlessness of my design to be a writer in spite of mother and hell" (W140), and that meant that a strong male writer would choose literature over journalism, individual effort over collaboration. Hence his decisions to work alone, yet the process of writing this novel proved to be mysterious and often humbling.

For in a large measure the book grew by itself, independent of his will or control. His creative process was strongly visual, aided by pictures and the work diary, which shifted him from a narrow "I" back to a broad, encompassing "We," the source of epic narrative. In the end, Steinbeck came to see that his creativity rose from both internal and external forces, not all of which he could control. His scenes of procreation and nurture bear the subtle imprint of authorial reflection. He had not yet fathered children himself, but Rose of Sharon became his opportunity to imagine birth through another, one of the opposite gender. That final image of a mother nursing an old man ends the novel appropriately. As Steinbeck told his editor, the closing incident "must be an accident, it must be a stranger, and it must be quick. . . . The fact that the Joads don't know him, don't care about him, have no ties to him—that is the emphasis. The giving of the breast has no more sentiment than the giving of a piece of bread" (L178). The anonymous stranger feeds to survive, just as his author succeeds in delivering a story "which is so much greater than I" (W36), fiction that melded idea and fact, invention and reporting. That recognition suggests that *The Grapes of Wrath* endures as literature because it sprang from journalism, a strong and vibrant mother.

NOTES

Parenthetical and abbreviated references are to:

B   Jackson J. Benson, *The True Adventures of John Steinbeck, Writer* (New York: Viking, 1984)
L   *Steinbeck: A Life in Letters,* Elaine Steinbeck and Robert Wallsten, eds. (New York: Viking, 1975)
W   *Working Days: The Journals of the Grapes of Wrath,* Robert DeMott, ed. (New York: Viking, 1989)
    John Steinbeck, *The Grapes of Wrath:* Text and Criticism, Peter Lisca, ed. (New York: Viking, 1972)

1. A summary of critical responses to the last scene appears in Jules Chametzky, "The Ambivalent Endings of *The Grapes of Wrath," Modern Fiction Studies* 11:34–44 (Spring, 1965).

2. The best survey of Steinbeck as social novelist is David P. Peeler, *Hope Among Us Yet* (Athens: University of Georgia Press, 1987), pp. 156–65. For his place in aesthetic tradition, see John H. Timmerman, *John Steinbeck's Fiction* (Norman: University of Oklahoma Press, 1986), pp. 3–41.

3. Other references appear throughout Robert DeMott, *Steinbeck's Reading* (New York: Garland, 1984) and Tetsumaro Hayashi, *A New Steinbeck Bibliography* (Metuchen: The Scarecrow Press, 1973).

4. Joseph Henry Jackson, "Why Steinbeck Wrote *The Grapes of Wrath," Booklets for Bookmen* 1:8–10 (1940). For summaries of the novel's reception see Peter Lisca's edition of *The Grapes of Wrath,* pp. 695–707, and David Wyatt's introduction to *New Essays on* THE GRAPES OF WRATH (Cambridge: Cambridge University Press, 1990). James Boylan, "Publicity for the Great Depression," in *Mass Media Between the Wars,* Catherine L. Covert and John D. Stevens, eds. (Syracuse: Syracuse University Press, 1984), p. 170.

5. William Stott, *Documentary Expression and Thirties America.* (New York: Oxford University Press, 1973), p. 119; see also Alfred Kazin, *On Native Grounds* (New York: Harcourt, Brace, 1942, rpt. 1956), pp. 382–98; and Karin Ohrn, *Dorothea Lange and the Documentary Tradition* (Baton Rouge: Louisiana State University Press, 1980). For a poststructuralist concept of documentary fiction, see Barbara Foley, *Telling the Truth* (Ithaca: Cornell University Press, 1986), a book that unaccountably excludes Steinbeck.

Considerations of post-1930s literary journalism appear in Ronald Weber, *The Reporter as Artist* (New York: Hastings House, 1974); *The John McPhee Reader,* William Howarth, ed. (New York: Farrar, Straus, & Giroux, 1976); John Hollowell, *Fact and Fiction* (Chapel Hill: University of North Carolina Press, 1977); Ronald

Weber, *The Literature of Fact* (Athens: Ohio University Press, 1980); John Hellman, *Fables of Fact* (Urbana: University of Illinois Press, 1981); *The Literary Journalists*, Norman Sims, ed. (New York: Ballantine Books, 1984); Shelley Fisher Fishkin, *From Fact to Fiction* (Baltimore: Johns Hopkins University Press, 1985); Howard Good, *Acquainted with the Night* (Metuchen: Scarecrow Press, 1986); and Chris Anderson, *Style as Argument* (Carbondale: Southern Illinois University Press, 1987).

6. Often called "intercalary" or "inter" chapters, but Steinbeck's terms "general" and "particular" are clearer. See Lisca's edition of *The Grapes of Wrath*, pp. 731–32. While critics have long known that Steinbeck used documentary sources, they assume such materials mainly affected his "general" chapters. For an example, see Joseph Fontenrose, *John Steinbeck* (New York: Barnes & Noble, 1963), p. 69.

7. Some of Steinbeck's antibourgeois attitudes probably expressed his rebellion against a middle-class background. See James Woodress, "John Steinbeck: Hostage to Fortune," *South Atlantic Quarterly* LXIII:386 (Summer, 1964).

8. Bristol's recollections and pictures appear in *Steinbeck* (Fall, 1988), pp. 6–8; David Roberts, "Travels with Steinbeck," *American Photographer* 22:45 (March, 1989); and Jack Kelley, "Travels with Steinbeck," *People* (April 24, 1989), pp. 67–74. The articles all differ on key points of evidence and emphasis.

9. Kelley, p. 73. Time, Inc., did not ignore the California labor migrants; see the unsigned "I Wonder Where We Can Go Now," *Fortune* (April, 1939), reprinted in the Lisca edition, pp. 625–42. This forceful, well-researched article appeared just before publication of *The Grapes of Wrath*.

10. Kelley, p. 73.

11. Reporting accurately on migrants was difficult because they were transient and wary of strangers. Steinbeck also faced the ethical problem of writing about poverty without patronizing its victims. For a moving statement on this quandary, see Robert Coles, "James Agee's Search," *Raritan* (Summer, 1983), pp. 74–100.

12. Lange's pictures—of migrant families, roadside scenes, and government camps—appeared in all but the seventh installment, which contained Steinbeck's stern denunciation of "Fascism in California." A full account of the Steinbeck–Collins relationship is Jackson J. Benson, "To Tom, who Lived it: John Steinbeck and the Man from Weedpatch," *Journal of Modern Literature* (April, 1976), pp. 151–94. See also B359–63.

13. Stott, pp. 58–63. For background on FSA photography, see Jack F. Hurley, *Portrait of a Decade* (Baton Rouge: Louisiana State University Press, 1972); Penelope Dixon, *Photographers of the Farm Security Administration* (New York: Garland, 1983); and Pete Daniel, *Official Images* (Washington: Smithsonian Institution, 1987). Studies of Lange include Karin Ohrn, *Dorothea Lange and the Documentary Tradition* (Baton Rouge: Louisiana State University Press, 1980); Jefferson Hunter, *Image and Word* (Cambridge: Harvard University Press, 1987), pp. 88–102; and Carol Schloss, *In Visible Light* (New York: Oxford University Press, 1987). Steinbeck wrote affectionately to Lange not long before her death in 1965; see *Steinbeck* (Summer, 1989), pp. 6–7.

14. Stott, p. 59; Wendy Kozol, "Madonnas of the Fields: Photography, Gender, and 1930s Farm Relief," *Genders* 2:17 (July 1988). See also "Photographic Contriv-

ance" in Hunter, pp. 99–105. *The Blood is Strong* closed with an epilogue, dated "Spring—1938", that described Steinbeck's recent experiences at Visalia: "And then in the rains, with insufficient food, the children develop colds because the ground in the tents is wet. . . . I talked to a girl with a baby and offered her a cigarette. She took two puffs and vomited in the street. She was ashamed. She shouldn't have tried to smoke, she said[,] for she hadn't eaten for two days. I heard a man whimpering that the baby was sucking but nothing came out of the breast. . . . Must the hunger become anger and the anger fury before anything will be done?"

15. Joseph R. Millichap, *Steinbeck and Film* (New York: Ungar, 1983), p. 29.

16. For a partial reprint of *Life* (June 5, 1939) see Arthur P. Moella, *FDR: The Intimate Presidency* (Washington: Smithsonian Institution, 1982), p. 64.

17. Mark Twain's epigram appears in *Following the Equator* (1897), as the head-note to Chapter 15. For the genres of documentary, see John Puckett, *Five Photo-Textual Documentaries from the Great Depression* (Ann Arbor: UMI Research Press, 1984); also Stott on "Documentary NonFiction," pp. 141–238; Hunter on "Collaborations," pp. 33–64; and Peeler on "Traveling Reporters," pp. 13–56.

18. David Wyatt suggests that ownership eventually became a divisive force in Steinbeck's first marriage. See *The Fall into Eden* (Cambridge: Cambridge University Press, 1986), p. 154. Steinbeck was no slouch at business; when looking for a new publisher, he told his agent "Get a [Dun and] Bradstreet report on whoever you pick. All things else being equal, pick the one who makes the highest offer" (L169).

19. Susan Sontag, *On Photography* (New York: Farrar, Straus & Giroux, 1978), p. 69. Years later, Bristol identified this portrait as the model for Rose of Sharon, "the character we had photographed together in the flooded boxcars of Visalia." See Bristol, p. 6.

20. Ulrich Keller, *The Highway as Habitat* (Santa Barbara: University Art Museum, 1986), pp. 29–32.

21. Millichap, pp. 1–7, summarizes Steinbeck's relationship to 1930s films.

22. Pare Lorentz, *The River* (New York: Stackpole Sons, 1938). This published text, subtitled "A National Drama in Pictures and Sound," includes many still photographs from FSA files.

23. Steinbeck was often attacked for *not* having described typical events or people, however. See Frank J. Taylor, "California's *Grapes of Wrath*," *Forum* 102:232–38 (November, 1939); reprinted in the Lisca edition, pp. 643–56.

24. Millichap, pp. 32–38.

25. McWilliams defended Steinbeck as an accurate reporter in "California Pastoral," *Antioch Review* 2:103–21 (March, 1942); reprinted in the Lisca edition, pp. 657–79.

26. *The Forgotten Village* (New York: Viking Press, 1941), pp. 6–7.

4

Joseph Mitchell and
The New Yorker Nonfiction Writers

Norman Sims

At *The New Yorker,* when you get off the elevator you step into an off-white, narrow little prison of a waiting room. The receptionist phones the inner sanctum of the editorial offices, and your host meets you at the door.

My host was Joseph Mitchell, who has been with *The New Yorker* since 1938. Although he was eighty-one-years-old and rumored to be a ghostly presence in the corridors of the magazine, he carried the grace of a much younger man. Mitchell's last magazine article appeared in 1964. He has regularly gone to his office since then, feeding the speculation that this very private man has been writing some magnificent addition to the books he published between 1938 and 1965.[1] Curiosity has been fed by Mitchell's own last work, *Joe Gould's Secret,* and by the appearance of a character similar to him in Jay McInerney's *Bright Lights, Big City.* Not surprisingly, given his longevity at the institution, his office is the first one down the hallway. The furnishings—metal desk, cabinets, flooring that dates from the age of linoleum—are standard at *The New Yorker.* That narrow office was a place I never expected to reach. For years, Mitchell has turned down requests for interviews. Finding out what he has been doing the last twenty-five years was the least of my objectives. There are deeper mysteries in his writing.

Mitchell and several of his colleagues at *The New Yorker* were responsible for keeping literary journalism alive during the middle years of the twentieth century before the New Journalism burst on the American scene. Mitchell's nonfiction, and to some extent that of A. J. Liebling, adopted a creative approach and probed further into the borderlands of fiction and nonfiction than did many of the highly publicized experiments of the New Journalism.

In the years just before World War II, *The New Yorker* magazine began nurturing literary journalists. Harold Ross founded *The New Yorker* in 1925 as a magazine dedicated to humor, criticism, short fiction, and reportage. Its early success owed little to literary journalism, and a great deal to a talented staff. Katharine Angell edited the fiction department while E. B. White, later her husband, in his "Notes and Comment" essays developed a voice that would be called "*The New Yorker* style." James Thurber contributed short pieces and humorous drawings that cemented both his reputation and the magazine's.

The genius who created *The New Yorker* was not necessarily a genius for organization. Ross fumbled repeatedly while looking for a managing editor. In 1933, he finally hired an editor who could make sense of his editorial system, and who would make a difference in the future of literary journalism. William Shawn arrived as a "Talk of the Town" writer and by 1939 was managing editor. After Ross's death in 1951, Shawn succeeded him as editor and served for thirty-five years.

Shawn's rise to power came at an opportune time. The *New Yorker* editorial corps had weakened as three of its foundation stones departed. Thurber's eyesight was failing and he steadily withdrew. Katharine and E. B. White moved to Maine in 1938, temporarily depriving the magazine of their guidance and contributions. The vacuum was gradually filled by new writers who made enduring contributions to literary journalism: John Jersey, John McNulty, Geoffrey Hellman, Joel Sayre, Alva Johnston, St. Clair McKelway, Philip Hamburger, John Lardner, Brendan Gill, Berton Roueché, John Bainbridge, and Lillian Ross. Referring to himself, Joe Mitchell, Jack Alexander, Richard O. Boyer, and Meyer Berger, A. J. Liebling once wrote, "I still think *The New Yorker*'s reporting before we got on it was pretty shoddy."[2]

Before they came to the magazine, most of the new writers had been newspaper feature reporters. Feature writing could be creative, especially under the editorship of someone like Stanley Walker, city editor of the New York *Herald Tribune,* but was severely limited in the time spent reporting and the scope of presentation. Moving to *The New Yorker* gave writers more time to work, more space (in print, if not in their cubbyhole offices), superb editing, greater autonomy, and—at least in the cases of Mitchell and Liebling—opportunities to pursue literary goals. The institutional conditions were ripe for literary journalism. Until the late 1950s, magazine writing had not fully exploited storytelling; one student of the era found little use of scenes, dramatization, or first-person narrative outside of *The New Yorker.*[3] No American magazine had offered the consistent freedom and encouragement found at *The New Yorker.* The payoff came rapidly from writers such as Mitchell, Liebling, Hersey, Boyer, and Lillian Ross.

Mitchell is a bright-eyed, energetic man who puzzles over things and takes pains to get them right. He dresses as he writes, in a stylish, comfortable, yet precise manner. He lacks a striking physical feature and never intrudes abruptly on a conversation. Talking with him is easy. His courtesy may be his most distinctive trait, along with an incredible memory. He grew up in the cotton and tobacco region near Fairmont, North Carolina, where his ancestors had lived since before the Revolutionary War. After four years at the University of North Carolina, he became a reporter for the New York *Herald Tribune.* He worked at the *Herald Tribune* and *World-Telegram* until 1938, except, as he said, "for a period in 1931 when I got sick of the whole business and went to sea."[4] Thereafter, he wrote profiles for *The New Yorker* of waterfront workers, people on the Bowery, Mohawk Indians who work on high structural steel, and characters from the Fulton Fish Market in the southeast corner of Manhattan near the Brooklyn Bridge. The literary critic Stanley Edgar Hyman put Mitchell in the tradition of William Faulkner, Saul Bellow, and James Joyce. Hyman said Mitchell "is a reporter only in the sense that Defoe is a reporter."[5] Like Defoe in *A Journal of the Plague Year,* Mitchell wrote articles that are mixtures of fiction and nonfiction. "Mr. Flood," a ninety-three-year-old retired house-wrecking contractor who lived in a waterfront hotel and pursued his remaining ambition of eating fish every day (and practically nothing else) and thereby living to be 115, was a composite character. "Combined in him are aspects of several old men who work or hang out in Fulton Fish Market, or who did in the past," Mitchell explained. "I wanted these stories to be truthful rather than factual, but they are solidly based on facts."[6]

Two of Mitchell's books illustrate the advances in literary journalism at *The New Yorker* from the late 1930s to the 1960s. In *The Bottom of the Harbor,* Mitchell reprinted magazine pieces written between 1944 and 1959, including "Up in the Old Hotel," his symbolic cultural portrait of the Fulton Fish Market. This piece illustrates Malcolm Cowley's remark that "Mitchell . . . likes to start with an unimportant hero, but he collects all the facts about him, arranges them to give the desired effects, and usually ends by describing the customs of a whole community."[7] *The Bottom of the Harbor* also contained "The Rivermen" and "Mr. Hunter's Grave," examples of Mitchell at his best in searching out the psychological core of a person or the symbolic meaning of a topic. Mitchell's last book, *Joe Gould's Secret,* published in 1965 at the dawn of the New Journalism, represents self-expression in nonfiction that stands somewhere between the realist and modernist styles found among New Journalists.

From the time Mitchell began writing about the Fulton Fish Market, he had a vision of a book that might report on the complexities of the characters he found there. He thought of writing about the fish market in the same way Melville wrote about whaling in *Moby Dick.* "I had an idea for a big book on the fish market," Mitchell said.[8] "I had those reefer trucks coming in from the East Coast, the West Coast, and the Gulf Coast and converging at South Street and Fulton Street early in the morning. I worked on it for years, and I couldn't find a focus or a way of telling it. I was lost out there among the bins of fish, screaming 'Rescue me!' Louie rescued me."

Louie Morino owned a seafood restaurant at 92 South Street, which still exists today in the South Street Seaport development around the Fulton Fish Market. The restaurant, Sloppy Louie's, occupied the first floor of a six-story building that was once a hotel. Mitchell began his story, "Up in the Old Hotel," with a description of the early-morning scene in the fish market, forty to sixty kinds of fish arriving from all over. Louie was introduced as the son of a northern Italian fisherman, Giuseppe Morino. His family fished from the village of Recco, near Genoa, from Roman times. His father specialized in octopus, enjoying exclusive rights to fish an underwater cave "full of octopuses; it was choked with them." In a few hours of bobbing meat scraps near the dark entrance of the cave, he could catch enough to "glut the market in Recco." Louie left Recco in 1905 at the age of seventeen. In America, he worked in restaurants and, by then in his mid-sixties, bought the run-down restaurant in the fish market. Louie's story became a vehicle for Mitchell's cultural portrait. He lingered over the customers who frequented Sloppy Louie's, their businesses, their attitudes, their habits, and their characteristic jokes.

Louie used the second floor of the building for storage, but the floors beyond were a mystery to him. The stairs only reached the second floor. An old elevator on a rope-pull then led upward. Louie did not want to risk a ride in the creaky old cage. "It makes me uneasy," Louie said, "all closed in, and all that furry dust. It makes me think of a coffin, the inside of a coffin. Either that or a cave, the mouth of a cave." After Louie explained the history of the building, from its days as the Fulton Ferry Hotel before the Brooklyn Bridge was built nearby, Mitchell said, "Look, Louie, I'll go up in the elevator with you."

The reader is amply prepared for some sort of symbolic climax. The building had been owned by the Schermerhorns, an important Dutch family of Old New York, and the hotel was near busy oceanic and coastwise steamship piers. Louie thought they might find "beds and bureaus, pitchers and bowls, chamber pots, mirrors, brass spittoons, odds and ends, old hotel registers that the rats chew on to get paper to line their nests with, God knows what all."

Louie brought out construction helmets and flashlights, and he and Mitchell climbed into the elevator. "Oh, God in Heaven," Louie cried, "the dust in here! It's like somebody emptied a vacuum-cleaner bag in here." Louis pulled them up to the third floor, and they stepped out into what was once the reading room of the Fulton Ferry Hotel. After the build-up, the discoveries were a disappointment. They found a bit of old hotel junk under a deep layer of dust: bedsprings, a tin water-cooler, a cracked glass bell, rusted sugar bowls, and a wire basket filled with empty whiskey bottles. A drawer in a bureau held "a few hairpins, and some buttons, and a comb with several teeth missing, and a needle with a bit of black thread in its eye, and a scattering of worn playing cards." The other drawers were empty. In the next bureau, Louie found a medicine bottle with "two inches of colorless liquid and half an inch of black sediment." He opened it and smelled. "It's gone dead. It doesn't smell like anything at all," Louie said. Then the restaurant owner grew morose and insisted they return to the ground floor. On the way down he said, "I didn't learn much I didn't know before."

"This climax is a tremendous let-down, and it is meant to be," Noel Perrin wrote in a retrospective article on Mitchell.[9] "They have broken through to the past, and all they find is trivial debris. For once the past had seemed retrievable—but when you reach out to seize it, you find nothing but dust and decay." Stanley Edgar Hyman took a more symbolist view of the several stories in *The Bottom of the Harbor,* a view encouraged by the literary perspectives of Kenneth Burke. Hyman wrote:

> Mitchell's other major theme, most boldly imaged in *The Bottom of the Harbor,* is the depths of the unconscious. Dusty hotel rooms shut up for decades and now reluctantly explored are infantile experience; the wrecks on the bottom of the harbor, teeming with marine life, are festering failures and guilts; the rats that come boldly out of their holes in the dark before dawn are Id wishes; Mr. Poole's dream of the draining of New York harbor by earthquake is a paradigm of psychoanalysis; Mr. Hunter's grave is at once tomb and womb and marriage bed.[10]

Such interpretations are uncommon in nonfiction, a literary realm where symbolism is often considered nonexistent or accidental. In one interview, Mitchell explained the symbolic significance of the story as he saw it. "Louie was always talking about his father, who was a fisherman. He had some underwater caves down there where octopuses lived. The first title I had on that story was 'The Cave.' I realized later on that upstairs in this old hotel was Louie going into the past, into the cave of the past. An octopus is a dream creature itself. It comes out of a nightmare sort of world. Later on, Stanley Hyman wrote a piece about this and I said, 'My God,

Stanley's right.' These were the nightmare figures upstairs of Louie's and maybe of mine. I didn't go out consciously looking for this, but maybe in some unconscious way I knew what I was doing."

The symbolism of the past, and Mitchell's peculiar interest in graveyard humor, played a part in two other articles, "The Rivermen" and "Mr. Hunter's Grave." Mitchell said Harold Ross, the founder of *The New Yorker,* came into his office one time to discuss the characters in Mitchell's articles. "You know," Ross said, "you're a pretty gloomy guy." A moment later, Ross amplified this remark. "Of course," he continued, "I'm no Goddamned little ray of sunshine myself." The concept of graveyard humor is "the only view I have of the world," Mitchell told me. "By that time I was writing about disappointed old men and old women. When they started talking about how nothing had turned out the way they thought it would, I said to myself, 'I can respond to that.' "

Mitchell reached into a drawer of his desk and took out a book of engravings by the Mexican artist José Guadalupe Posada. On the cover was an engraving of a laughing skeleton playing a guitar. "This strange man Posada has had a great influence on me—on the way I look at the world," Mitchell said. "I first heard of him in 1933, during the worst days of the Depression, when I was a reporter on the *World-Telegram.* I had gone up to the Barbizon Plaza Hotel to interview Frida Kahlo, who was the wife of Diego Rivera and a great painter herself, a sort of demonic surrealist—I believe time might tell that she is a greater painter than Rivera. That was when Rivera was doing those Rockefeller Center murals. Thumb-tacked all over the walls of the hotel suite were some very odd engravings printed on the cheapest kind of newsprint. 'José Guadalupe Posada,' Kahlo said, almost reverentially. 'Mexican. 1852–1913.' Then, through an interpreter, she told me that she had tacked them up herself so she could glance at them now and then and keep her sanity while living in New York City. Some were broadsides. 'They show sensational happenings that took place in Mexico City in streets and in markets and in churches and in bedrooms,' Kahlo said, 'and they were sold on the streets by peddlers for pennies.' One broadside showed a streetcar that had struck a hearse and had knocked the coffin onto the tracks. A distinguished-looking man lay in the ruins of the coffin, flat on his back, his hands folded. One showed a priest who had hung himself in a cathedral. One showed a man on his deathbed at the moment when his soul was separating from his body. But the majority of the engravings were of animated skeletons mimicking living human beings engaged in many kinds of human activities, mimicking them and mocking them: a skeleton man on bended knee singing a love song to a skeleton woman, a skeleton man stepping into a confession box, skeletons at a wedding, skeletons at a funeral, skeletons making speeches, skeleton gentlemen in top hats, skeleton ladies in fashionable bonnets. But the most astonishing thing is that all these pictures were humorous, even the most morbid of them, even the busted coffin on the streetcar tracks. That is, they had a strong undercurrent of humor. It was the kind of humor that the old Dutch masters caught in those prints that show a miser locked in his room counting his money and Death is standing just outside the door. It was Old Testament humor, if I make any sense: the humor of Ecclesiastes—vanity, vanity, all is vanity. Gogolian humor. Brueghelian humor. I am thinking of that painting by Brueghel showing the halt leading the blind, which, as I see it, is graveyard humor. Anyway, ever since that afternoon in Frida Kahlo's hotel suite, I have been looking for books showing Posada engravings. I never pass a bookstore in a Spanish neighborhood of the city without going in and seeing if they have a Posada book, and I have found quite a few of them. My respect for him grows all the time. To me, he is the great master of what I think of as graveyard humor."

"The Rivermen" focused on the town of Edgewater, New Jersey, on the Hudson River across from the upper West Side of Manhattan within sight of the George Washington Bridge. Generation after generation of shad fishermen and rivermen have lived in Edgewater. As was his habit, Mitchell frequented the Edgewater Cemetery. From North Carolina, he brought an interest in wildflowers, which could be found most easily in overgrown cemeteries around New York City. At Edgewater, he wrote:

> Old men and old women come in the spring, with hoes and rakes, and clean off their family plots and plant old-fashioned flowers on them. Hollyhocks are widespread. Asparagus has been planted here and there, for its feathery, ferny sprays. One woman plants sunflow-

ers. Coarse, knotty, densely tangled rosebushes grow on several plots, hiding graves and gravestones. The roses that they produce are small and fragile and extraordinarily fragrant, and have waxy red hips almost as big as crab apples. Once, walking through the cemetery, I stopped and talked with an old woman who was down on her knees in her family plot, setting out some bulbs at the foot of a grave, and she remarked on the age of the rosebushes. "I believe some of the ones in here now were in here when I was a young woman, and I am past eighty," she said. "My mother—this is her grave—used to say there were rosebushes just like these all over this section when she was a girl. . . . And she said *her* mother—that's her grave over there—told her she had heard from *her* mother that all of them were descended from one bush that some poor uprooted woman who came to this country back in the Dutch times potted up and brought along with her . . . She thought they were a nuisance. All the same, for some reason of her own, she admired them, and enjoyed looking at them. 'I know why they do so well in here,' she'd say. 'They've got good strong roots that go right down into the graves.' "[11]

The image of ancient roses with their roots in the graves perfectly represents Mitchell's approach to reporting. It connects him to the symbolic manner of James Joyce—his favorite author—and to his own past. "On Sunday afternoons, when I was a child," he recalled, "my father and mother and my brothers and sisters would go for a ride in the country, and more often than not we would wind up visiting some old family cemetery out in a field in a grove of cedars where people who were kin to us were buried, and my parents would describe who this one was and who that one was and exactly how they were related to us, and somehow I always enjoyed this. And at least once every summer, my family and my two aunts on my mother's side and their families would meet on a Sunday afternoon at an old Scottish Presbyterian church out in the country, old Iona Church that my mother's family long ago had helped build. We would all bring some big watermelons from our own gardens—big long green Rattlesnakes and big round Cuban Queens—and we would sit the melons on tables in a picnic grove in back of the church and slice them into rashers and eat them, and then we would stand around and talk, and then, late in the afternoon, as it was getting dark, we would go for a walk up and down the rows in the cemetery that belongs to the church and my mother and my aunts would comment on the people buried there, just about all of whom were related to them. 'This man buried here was so *mean,*' one of them would say, 'I don't know how his family stood him.' And a few steps farther along, she would say, 'And this one here was so *good,* I don't know how his family stood him.' Every time I read the Anna Livia Plurabelle section in *Finnegans Wake,* I can hear the voices of my aunts as they walk among the graves in Iona cemetery and the sun is going down.

"And there's another thing I remember about cemeteries when I was a child. In those days, the Baptists in Fairmont used to have a big Easter-egg hunt in the old cemetery across the road from the First Baptist Church. My father's people were Baptists and some of my ancestors, including my paternal grandfather, were buried in the cemetery, and I always went to the Easter-egg hunt. The Sunday-school teachers would hide colored eggs all over the cemetery, in among the graves. Then, later in the day, the children would line up, and, at a signal, start running all over the place, hunting for the eggs. I can still remember finding a beautiful robin's-egg-blue egg under a pile of dead leaves on my grandfather's grave."

The community of the rivermen in Edgewater had survived for generations by working on tugboats and excursion boats on the river, by fishing for shad during the annual run, and before that by cutting paving blocks for New York City from a local quarry. "In the years when there was no shad fishing going on, you could walk along there and you would have no idea—unless you knew what the barges represented—that this thing went on. There was a kind of secret background, but there are traces all along if you happen to know it," Mitchell said. He was trying to preserve that past in his story, perhaps as a seed of resurrection, an egg in a cemetery.

Mitchell found similar traditions in the Fulton Fish Market, where the companies would retain the names such as "Chesebro Brothers, Robbins and Graham" long after the Chesebro brothers and Robbins and Graham were all dead and gone. "Whatever quality is in my newspaper or magazine writing has come from the desire to put in a background, which is constantly changing, people who are constantly changing, but who display an *attitude* toward life and death that doesn't change," Mitchell said. The woman tending roses in Edgewater had the enduring attitude that attracted Mitchell.

In "Mr. Hunter's Grave," Mitchell profiled an elderly black bricklayer, a resident of a traditionally black community along Bloomingdale Road on Staten Island. The story begins, innocently enough, as a quest for wildflowers in a Staten Island cemetery. Mitchell was directed to Mr. Hunter, who knew the location of the cemetery, but before long he was listening to the story of a man's life and drawing out of it the culture of the entire community.

"My whole idea of reporting—particularly reporting on conversation—is to talk to a man or a woman long enough under different circumstances, like old Mr. Hunter down on Staten Island, until, in effect, they reveal their inner selves," he said. "During the time that I knew him, Mr. Hunter's son died—his son was an alcoholic—and I genuinely sympathized with him, and he gradually saw that I was seriously and deeply interested in the life of his community. He himself was trying to understand what his mother went through bringing up her family. I was always trying to reach his inner life. I can't really write about anybody until they speak what I consider 'the revealing remark,' or the revealing anecdote or the thing that touched them. I've often deliberately tried to find those things. You're trying to report, at the beginning without knowing it, the unconscious as well as the consciousness of a man or woman. Once I had what I considered the revealing remark, I could use that to encourage them to talk more about that aspect of their lives. They were able to talk, like Mr. Hunter could talk about his first wife's death, about his son's death, about his stepfather who he hated and who I guess hated him. That way I could go far deeper into the man's life than I could any other way."

This approach led Mitchell toward the use of symbolic backgrounds and a technique he attributes to T. S. Eliot. "One thing you have to do, if you're going to write this sort of thing, is realize that people have buried their pain and have transformed experience enough to allow them to endure it and bear it. If you stay with them long enough, you let them reveal the past to themselves, thereby revealing it to you. Then they will dare to bring out the truth of something even if it makes them look bad."

Mitchell's technique departed from the standard nonfiction of the day. While most magazine articles presented an almost flat character defined by the facts of age, occupation and achievements, Mitchell's writing took the reader into a character's inner life. With Mr. Hunter, the revealing moment came as he told Mitchell about a time when he saw his mother's reflection in the window of a store. As she passed by in the street, Mr. Hunter saw her face and how sad it was. The reflection made the familiar strange, and Mr. Hunter could thereby interpret it. Mitchell said he wanted to create those backgrounds in his nonfiction writing, producing an experience that we are capable of understanding more clearly than we can in everyday life. Writing about moments that reveal an inner life demands an approach different from the frontal assault of standard journalism. "T. S. Eliot called it the 'objective correlative,' " Mitchell said. "It's where you write about one thing and you're actually writing about another. Or where you make one thing represent another." He cited, as an example, a passage in D. H. Lawrence's *Sea and Sardinia* in which a man is roasting a young goat at a large, open fireplace. The scene presents a primitively spitted goat being roasted by an Italian peasant, greasy from the work he is doing, against the flickering shadows cast by the open flames and the bluish glow of burning fat. "Suddenly you begin to realize," Mitchell said, "that Lawrence is showing you how it was in the caves. He doesn't anywhere imply that, but you can't miss it.

"In other words, first you write the background for something to happen, and then it happens against this background. It has to be in the round. You have to have the person against the background. That gives a story meaning and significance, rather than plot."

Joe Gould's Secret was a two-part profile for *The New Yorker,* with the parts separated by twenty-two years. Mitchell met Joe Gould in Greenwich Village in 1938. Gould was a bohemian, a short, gaunt, garrulous man with a disheveled beard and a wild look. He was the son of a New England doctor and a graduate of Harvard, class of 1911. His ancestors had been in the New World since 1635. On the Bowery, Gould wore castoff clothes and he cadged drinks by demonstrating how he conversed with sea gulls.

Gould was always scribbling in nickel composition books, writing the Oral History of Our Time, which had grown over the years to fill hundreds of notebooks with millions of words. Gould's project, which had been reported by writers on several publications, was the focus of the first profile. The Oral History contained ordinary conversations, biographies of bums, sailor's tales, hospital experiences, harangues from speakers in Union Square and Columbus Circle, dirty stories, graffiti found in washrooms, gossip, accounts of Greenwich Village parties, and arguments about topics of the day such as free love, birth control, psychoanalysis, Christian Science, alcoholism, and art. Gould's inspiration came from William Butler Yeats, who once commented, "The history of a nation is not in parliaments and battlefields, but in what the people say to each other on fair days and high days, and in how they farm, and quarrel, and go on pilgrimage." Gould explained to Mitchell:

> All at once, the idea for the Oral History occurred to me: I would spend the rest of my life going about the city listening to people— eavesdropping, if necessary—and writing down whatever I heard them say that sounded revealing to me, no matter how boring or idiotic or vulgar or obscene it might sound to others. I could see the whole thing in my mind—long-winded conversations and short and snappy conversations, brilliant conversations and foolish conversations, curses, catch phrases, coarse remarks, snatches of quarrels, the mutterings of drunks and crazy people, the entreaties of beggars and bums, the propositions of prostitutes, the spiels of pitchmen and peddlers, the sermons of street preachers, shouts in the night, wild rumors, cries from the heart.[12]

The idea of the Oral History attracted Mitchell. His own profiles had focused on ordinary people, especially on the Bowery and in the fish market—people who lived life fully but in ordinary ways. Describing five of the articles in *The Bottom of the Harbor,* Noel Perrin said, "each tells its story so much in the words of its characters that it feels like a kind of apotheosis of oral history."[13]

One problem flawed the first profile, "Professor Sea Gull," which Mitchell published in 1942: he never got a chance to read the Oral History, which had been reported as filling so many notebooks that they could be stacked higher than Gould himself. Gould said his notebooks were stored in friends' basements and attics, and the opportunity never arrived for Mitchell to read them.

In the 1964 profile, Mitchell reported his efforts to find the Oral History and read some portion of it. His search led to a few notebooks filled with repetitive and obsessive memories of Gould's father, who had once made a psychologically damaging remark about his son's chances in life. After considerable searching, Mitchell discovered Joe Gould's secret: the Oral History did not exist. Gould's convincing patter had seemed unassailable to several reporters, including Mitchell. After announcing his conclusion to Gould, Mitchell immediately regretted it.

> I returned to my office and sat down and propped my elbows on my desk and put my head in my hands. I have always deeply disliked seeing anyone shown up or found out or caught in a lie or caught red-handed doing anything, and now, with time to think things over, I began to feel ashamed of myself for the way I had lost my temper and pounced on Gould."[14]

Charitably, Mitchell decided the Oral History existed in Gould's head and might one day be written down on paper.

> It was easy for me to see how this could be, for it reminded me of a novel that I had once intended to write. I was twenty-four years old at the time and had just come under the spell of Joyce's "Ulysses." My novel was to be "about" New York City. It was also to be about a day and a night in the life of a young reporter in New York City. . . . But the truth is, I never actually wrote a word of it . . . When I thought of the cataracts of books, the Niagaras of books, the rushing rivers of books, the oceans of books, the tons and truckloads and trainloads of books that were pouring off the presses of the world at that moment, only a very few of which would be worth picking up and looking at, let alone reading, I began to feel that it was admirable that he *hadn't* written it.[15]

Later on, Mitchell wrote, he received a letter from a friend of Gould's. "I have always felt that the city's unconscious may be trying to speak to us through Joe Gould," the letter said. "And that the people who have gone underground in the city may be trying to speak to us through him. And that the city's living dead may be trying to speak to us through him. People who never belonged anyplace from the beginning. People sitting in those terrible dark barrooms."[16]

Stanley Edgar Hyman's essay, "The Art of Joseph Mitchell," presented this rather pregnant interpretation of the book:

> In literary terms, *Joe Gould's Secret* is a Jamesian story of life's necessary illusion (the secret is that the nine-million-word Oral

History of Our Time that Gould had spent his lifetime producing
did not exist). The book is written, however, not in intricate James-
ian prose, but in the bubbling, overflowing manner of James
Joyce . . .

    In deeper terms, Gould is a masking (and finally an unmasking)
for Mitchell himself. In *Joe Gould's Secret* Mitchell seems freer than
ever before to talk about himself, his resolutions and intentions, his
methods of reporting and writing, even his life . . . By the end of
the book, when he discovers Gould's secret, Mitchell becomes, not
Gould's bearer or Gould's victim, but Gould himself, and the unwritten Oral History merges with Mitchell's own unwritten novel, a
New York *Ulysses* (which Blooms magnificently even in four-page
synopsis). Then we realize that Gould has been Mitchell all along, a
misfit in a community of traditional occupations, statuses, and roles,
come to New York to express his special identity; finally we realize
that the body of Mitchell's work is precisely that Oral History of
Our Time that Gould himself could not write.[17]

Perhaps I took Hyman's thesis—that Gould is actually Mitchell—
too literally, but it brought together several streams that had been
running through my mind and merged them into a river I could not
ignore. For a while, I imagined that Gould never existed, and I
could see clues in the text.

In the first place, Mitchell had dealt with fictional characters
before, although they were certainly based on fact. His Old Mr.
Flood, the ninety-three-year-old resident of a hotel in the fish
market, was a composite of several characters in the market who
would not cooperate with Mitchell on a profile. Harold Ross suggested Mitchell write a composite as a profile, and a three-part
series followed. Mitchell told me that one part of Old Mr. Flood
was Joe Mitchell, the part of the character who only ate seafood.
"All the things I said in there about eating fish, that's what I
believe," Mitchell said. He called it a "seafoodetarian diet."

In addition, Gould says a lot of things about the Oral History
that sounded like the Joe Mitchell I had been interviewing. "Some
talk has an obvious meaning and nothing more, [Gould] said, and
some, often unbeknownst to the talker, has at least one other
meaning and sometimes several other meanings lurking around
inside its obvious meaning."[18] Gould says, "In autobiography and
biography, as in history, I have discovered, there are occasions
when the facts do not tell the truth."[19] Gould looks for the reveal-

ing remark in conversations, which was a hallmark of Mitchell's reporting technique. Mitchell would sit for many hours listening to someone like Mr. Hunter on Staten Island, waiting for the comments that cast light on a lifetime. "He was the sad old man you see in Balzac or Thomas Hardy," Mitchell said of Mr. Hunter, "the sad old figure in the corner someplace who wants to tell you what he learned going through life, but he never got the chance. To tell you the truth, after a while I got an idea that if I had any skill, it grew out of this fact that I'm not easily bored. I can listen indefinitely to anybody." It was also characteristic that Mitchell's real-life subjects were based on literary models.

Joe Gould's father, the doctor, had driven him away from home with his expectations. Joe Mitchell's father contributed to his move to New York City, although less harshly. Mitchell's father wanted him to become a cotton trader. Mitchell discovered while standing next to his father on the cotton-trading platform that he is almost a dyslectic in arithmetic. He knew he could never handle cut-throat cotton trading, so when he learned to write in college he took the opportunity to leave. "I always felt like an exile," he told me. So had Joe Gould.

Some passages in *Joe Gould's Secret* read like suggestive clues in a mystery. For example, one night a bohemian in Goody's bar said to Joe Gould, "You don't seem to be yourself." And Gould answered, "I'm *not* myself. I've never been myself."[20]

Last, the parallels are too close. What has Mitchell been doing the last twenty-five years, writing in Oral History? He has been in his office working on a book project that he says will collect all his work in one place. It will contain ali of his *New Yorker* articles, many of which were never reprinted, and some new work, such as a profile of Joe Cantalupo, who once hauled trash out of the fish market. This might prove to be his best writing, but no one has seen it. Joe Mitchell's secret might parallel Joe Gould's. Has he been not-writing nonfiction pieces, or perhaps his novel, or his own Oral History that would portray New York as Joyce portrayed Dublin? Has he been going through the motions while actually creating a mask? This possibility has occurred to virtually everyone who has commented on the Gould book.

It seemed a bit much to imagine that Joe Gould was really Joe Mitchell. But then, it was hard to imagine that the Oral History

did not exist. I never looked for Gould's birth certificate nor
evidence of his father's medical practice. I did confirm that Gould
was graduated from Harvard in 1911 and his father from Harvard
Medical School in 1888. Then I realized that such evidence did
not answer any questions. I believe Joe Gould really did exist, but
Joe Mitchell occupies a good portion of Gould's character, just as
he does Mr. Flood's. Whatever the nature of Gould's personality,
the symbolism of this book is pure Mitchell. Gould had created
his own identity, a mask that was as much an anchor for him as
anyone's more genuine activities. Mitchell always sought the true
story behind those masks, the window on the personality opened
by the revealing remark. He found the truth in Joe Gould's story,
but part of Joe Mitchell's life got entangled with it and created a
mystery.

The symbolism matters to the extent that it deepens our knowledge of Mitchell's work and of modern literary journalism. There
are mysteries in Joe Mitchell's prose that I could have asked him to
explain, although I knew he would have talked his way around
such questions as craftily as a politician. The mystery, after all,
gives the prose an embedded edge that I would hate to dull by
explaining it away. Was Mitchell really Joe Gould? Was Old Mr.
Flood an incomplete novel presented as a *New Yorker* nonfiction
profile? Were his shad fishermen and Fulton Fish Market types a
silent conspiracy by a very clever man to create an innovative
mixture of fiction and nonfiction, or were they merely results of
interviews with ordinary people? The clues in the texts could lead
to answers on both sides. He may have been Joe Gould. I did not
want to eliminate the delicious mystery because it was along that
edge of uncertainty that Mitchell's work soared toward greatness.

I did eventually ask him if his interests had mingled with Joe
Gould's to create this work. "Oh, Lord, yes," Mitchell said. "We
were in the same boat. We both came from small towns and didn't
fit in, and both had an idea. He had the same feeling about people
on the park bench talking. I was talking about myself here. He was
talking about himself and I was talking about myself."

"With all the people in New York City," I asked, "why does Joe
Gould become an interesting person to you?"

"Because he is me," Mitchell said. "God forgive me for my
version of Flaubert's remark about Madame Bovary. I think all of

us are divided up into lots of different aspects, you might say. To mix them up, you almost have to say, 'I am so-and-so,' just as I tried to do with Gould and all the different aspects of the people who had seen him."

Mitchell later elaborated. "Everything in the Gould book is documented, all those things in the *Dial* and all the records of his family. But I could have used this documentation in a different way. The creative aspect of it is the particularity of the facts that you choose, and the particularity of the conversations that you choose, and the fact that you stayed with the man long enough to get a panoply of conversations from which you can choose the ones that you decide are the most significant. The Gould I described, I think, is the absolutely true Gould. But another person could have written the story about Joe Gould far differently."

This sort of discovery seemed more important to me than finding out what he has been doing for the last twenty-five years. After saying he did not enjoy perpetuating mysteries, Mitchell finally explained. Whether or not he answered the question is another matter. His mother died in 1963. She left him two farms near Fairmont, his hometown, in Robeson County, North Carolina. He rented one and began managing the other in a partnership arrangement that required him to go to Robeson County often. "It so happened that farming was changing radically down there at that time," he said. "Many farmers were changing from the traditional crops—cotton and tobacco—to such crops as soybeans and winter wheat, and I became pretty much involved in that." Then, in 1976, his father died, at the age of ninety-five. "My father had been a farmer, a cotton buyer, a tobacco warehouseman, and a speculator in cotton futures," Mitchell said. "He had accumulated approximately 3,500 acres of land in tracts scattered all over the lower part of the county. He left this land to his six children—me, my two brothers, and my three sisters. Half of it was farmland and half of it was covered with timber, mostly tall, beautiful Southern short-leaf pines. State foresters who examined the timber told us that it was fully mature and was vulnerable to forest fires and pine beetles and that we should sell it, which we did. And as soon as the timber was cut, we started a reforestation program that is still going on. We are reforesting the cut-over land tract by tract, and I have become very interested in this. In fact, I guess I should say that I have become obsessively interested in it. In some periods of the year, I spend more time down in Robeson County working in the woods and staying in the old family house on Church Street in Fairmont than I do in New York City.

"Another thing that has interrupted my work is the South Street Seaport Museum. I was mixed up in the Seaport from the beginning—I was one of the charter members—and for eight years or so, from 1972 until 1980, I was a member of the restoration committee, which was the busiest of the Seaport committees, and that took up a great deal of my time. And then, in 1982, Mayor Koch appointed me to the New York Landmarks Preservation Commission and I spent five years as a commissioner. But despite all these interruptions, down South and up here, I have continued working on a book, and one of these days, if I am not terminally interrupted, I hope to finish it."

For many years, Mitchell avoided being interviewed because he had once turned down a friend, a newspaperman, who asked for an interview—he did not want to talk about inherited land and such—and he has tried to be consistent since then. In addition, he said, "As a journalist, I have the old feeling that the reporter should stay out of it." In the meantime, rumor and speculation have grown up around Mitchell, as it has around J. D. Salinger. Mitchell was not hiding and had no metaphysical objections to being interviewed; he simply felt until recently that he did not want to talk.

In *Joe Gould's Secret* and *Old Mr. Flood,* it seemed to me, Mitchell discovered an avenue for self-expression not often found in journalism. How does a writer inject himself into the narrative without upsetting readers who are accustomed to impersonal newspaper prose? Facing the same question, some writers, such as James Agee, threw themselves and their psyches into the foreground. On the other hand, Liebling portrayed himself as a secondary character along the margins of the storyline, especially in his early work. Writers could become a dominant part of the narrative, or stand apart from it and erase their personalities and motivations. Both choices were uncomfortable for creative nonfiction writers. Mitchell found another solution by merging himself with the characters of Mr. Flood and Joe Gould, and then writing about them in third person.

In 1952, Joe Liebling, writing about Colonel John R. Stingo ("The Honest Rainmaker"), "discovered a way to infuse himself into another man's nature in such a manner that he created an ostensibly real portrait that was in one sense fictional, in another, a retouched composite of two flesh-and-blood men," as Raymond Sokolov wrote.[21] Mitchell had already pioneered that ground. "I think Liebling's Col. Stingo is my Mr. Flood, to tell you the truth," Mitchell said.

Another attempt to merge with a central character can be seen in *Joe Gould's Secret,* but the work threatens to overwhelm the reader with mysteries about the legitimate links between fiction and nonfiction. Few nonfiction writers at that time had experimented so aggressively along the borders as did Liebling and Mitchell. Some that come to mind are Agee in *Let Us Now Praise Famous Men,* Ernest Hemingway's nonfiction reports on Spain and Africa, George Orwell in *Homage to Catalonia, Down and Out in Paris and London,* and *The Road to Wigan Pier,* and John Dos Passos's *Orient Express.* Mitchell is present in his work, as much as Norman Mailer or Joan Didion are in theirs, but modernist writers such as Didion and Mailer are self-consciously playing a role that the reader must interpret. Mitchell's presence remains hidden, perhaps even unconscious to him, and his motives and purposes remain set behind a veil of symbolism, his personality merged into that of another character. In several respects, this was territory not explored by the New Journalists; few others have moved to this borderland.

"You hope the reader won't be aware," Mitchell said. "If I read something and I think, 'Oh, God, here comes the myth,' I'm tired of it already. But if it's inherent and inescapable, then the reader will go along. You want to take the reader to the last sentence. I don't want to take him there just by *fact.* I want to take the reader there by going through an experience that I had that was revealing. There's something I like about that word 'reveal.' "

One way of looking at Joe Mitchell's work, I decided, was as a web of reporting, cultural anthropology, symbolism, and memoir. Old Mr. Flood and Joe Gould were made of a fabric that had once

clothed Joe Mitchell. At the same time, his writing opened the
door to the Fulton Fish Market, the Bowery, and Greenwich Village for readers who had no knowledge of those places, just as a
good cultural anthropologist can bring a Balinese village to life.
One of his strongest personal interests was literature, especially
James Joyce and the Russian literary greats. Whatever else was
going on in Mitchell's reporting, there was always the possibility of
an elaborately drawn symbolic meaning.

Of the several other *New Yorker* writers who were creating innovative literary journalism after the war—including most certainly
John Hersey and Lillian Ross—perhaps the most important was A.
J. Liebling. In 1935, Liebling moved to *The New Yorker,* completing a twelve-year newspaper career that began at age eighteen and
ended as a feature writer at the New York *World-Telegram,* where
he and Joe Mitchell became close friends. Shawn handled his copy
and his training from the beginning. From 1936 until he left for
Europe in September 1939, Liebling wrote profiles of lowlife street
characters, popular entertainers, con men, and boxers. Liebling
claimed he received the assignment to Europe on the eve of World
War II because he spoke French, but he was already a proven
writer—just the kind of person who would find a different approach to a spectacular story. While other reporters gobbled up
combat stories in Europe, Liebling sought out the French civilians
who had survived occupation by two armies, and the stories of
ordinary soldiers. After the war he retraced his route from a Normandy beachhead, where he had come ashore under fire, through
the farmhouses and restaurants he had visited on the road to Paris.
Liebling's sophisticated yet earthy first-person account stood a
head above contemporary journalism.[22] He was later best known
as *The New Yorker*'s press critic, and as the author of *Chicago: The
Second City* and *The Earl of Louisiana.*
   Liebling recognized in Mitchell a compatriot who had been drawing from the same models for his prose. They had both read
George Borrow's nineteenth-century books on the Spanish gypsies, François Villon, Robert Louis Stevenson, and Ben Hecht's
Erik Dorn. Liebling had read Baudelaire and Rimbaud in the
original. Both liked Stephen Crane, but Liebling preferred *Red
Badge of Courage,* while Mitchell, who grew up in the South but

could not stand Civil War stories and history, favored *Maggie: A Girl of the Streets*. They discussed writing by arguing about Stendhal's *The Red and the Black,* rather than by reviewing each other's journalism. "I didn't think of influencing Joe, or of Joe influencing me," Mitchell said. "We were too individualistic. We talked a lot about books but not much about our own writing. That was a private thing. We preferred to talk about how Stendhal did it." Both had read Rabelais, and Turgenev's *Sportsman's Sketches.* The list of models goes on: Sherwood Anderson's *Winesburg, Ohio;* Fielding's *Jonathan Wild; The Arabian Nights;* Kafka's short story "The Burrow"; Thomas Mann's "Disorder and Early Sorrow"; Robert Graves's *White Goddess;* the books of Mark Twain and James Joyce; John Skelton's poems. Liebling drew heavily on William Cobbett, Pierce Egan's boxing stories, Hazlitt, and Dostoevsky. These models are more literary than one might expect for nonfiction writers, but then, Mitchell and Liebling had extraordinary goals for their nonfiction. As Sokolov noted about Liebling's era and our own, "We cannot help noticing that we live in a time of confused genres, where energies focus at the no man's land between fact and fiction."[23]

The financial success of *The New Yorker* encouraged Mitchell, Liebling, and others who wanted to write literate reports about ordinary people, street eccentrics, and the cultures of the world of boxing or the Fulton Fish Market. Never generous with its writers, the magazine still found ways to support innovative journalism. Mark Singer, a contemporary *New Yorker* writer, has described the methods of compensation at the magazine as "Byzantine."[24] Originally, Harold Ross paid more for a celebrity "highlife" profile than for a "lowlife" piece. Mitchell said Ross added a new category to reward the newspaper feature writers in the postwar years. This was the "highlife lowlife" profile. "For example," Mitchell said, "in my Mazie profile [about a theater ticket-taker on the edge of the Bowery], Ross said, 'I'll tell you what we'll do. You've got Fannie Hurst in there. She's highlife. That makes it a "highlife lowlife." ' And then he put another classification in, that if it was a 'humorous highlife' it got a certain amount, but if it was a 'humorous highlife lowlife,' that was as high as you could get."

Liebling wrote first-person narratives, developing a master style rarely matched. John McNulty and Joe Mitchell created fictional or composite characters such as Mr. Flood, and brought them to life in the service of nonfiction. Of course, this was in an age before the controversies over the New Journalism, where composite characters such as Gail Sheehy's "Redpants" were roundly condemned not so much on theoretical grounds but more as a way of thumping the whole enterprise of New Journalism. Mitchell received little or no criticism for his literary experiments, and no excuses were needed. "Old Mr. Flood" worked and injured nobody. Even John Hersey used a composite character in "Joe Is Home Now," one of his World War II stories, although he would later criticise Tom Wolfe for similar "fiction-aping" practices.[25] Lillian Ross's career reached full stride just before the New Journalism took the stage. She wanted to stay out of her reports, yet she artfully created such a distinctive voice in her work that later New Journalists such as Sara Davidson would acknowledge her as a model they emulated.

In 1973, Tom Wolfe published an essay, "The New Journalism," in which he described how some reporters had paid homage to "The Novel." Even today, his article remains an oracle for scholars of the New Journalism. Back in the early 1960s, Wolfe said, the envious reporter with literary ambitions wanted to quit the newspaper game, write a novel of his own, and retire to a fishing shack in Arkansas. The literary scene accommodated journalists' dreams, he said, but only if they were dreams about fiction. "There was no such thing as a *literary* journalist working for popular magazines or newspapers. If a journalist aspired to literary status—then he had better have the sense and the courage to quit the popular press and try to get into the big league." In that atmosphere, Wolfe said a few journalists came up with "a curious new notion . . . in the nature of a discovery." This was the idea that "it just might be possible to write journalism that would . . . read like a novel."[26] Thus began the New Journalism—according to Tom Wolfe.

Wolfe's first-person account of his life as a feature writer in New York and the rise of New Journalism in the early 1960s carried such an air of authenticity that few have questioned it. But he was wrong when he said there was "no such thing as a *literary* journalist." He largely ignored the history of literary journalism at *The New Yorker,* the institution primarily responsible for the development of the form for thirty years.

In an appendix, Wolfe grudgingly mentioned John Hersey, Truman Capote, Lillian Ross, and A. J. Liebling as "Not Half-Bad Candidates" for historical forerunners to the New Journalism, grouping them with "various writers for *True.*" Wolfe's literary background, including a doctorate from Yale in American Studies and his tenure on the New York newspapers, suggests he should have heard about the work of Joe Mitchell, John McNulty, Meyer Berger, Alva Johnston, St. Clair McKelway, and the other *New Yorker* nonfiction writers. Why, then, did Tom Wolfe resist giving credit to *The New Yorker* writers when he drew his historical profile of the New Journalism? And why did the *New Yorker* writers, who are not shy, fail to defend their role in establishing the New Journalism?

For one thing, there was the "Tiny Mummies!" episode. This sordid little controversy created such a distaste for Wolfe that *New Yorker* writers avoided any association with the New Journalism, even when they deserved membership.

On April 11 and April 18, 1965, Wolfe published a two-part article in *New York* magazine, then the Sunday supplement of the *Herald Tribune.* The first of these, titled "Tiny Mummies! The True Story of The Ruler of 43d Street's Land of The Walking Dead!" attacked *The New Yorker*'s editor, William Shawn, as the "embalmer" of a dead institution.

Wolfe's articles raised the hackles of the literary community. Usually a reticent man, Shawn called the articles "false and libelous" and "wholly without precedent in respectable American journalism." Murray Kempton, Nat Hentoff, Walter Lippmann, and Joseph Alsop also expressed their displeasure. Several *New Yorker* writers rose in indignation to defend their editor. Renata Adler and Gerald Jonas of *The New Yorker* said Wolfe's articles were "reportorial incompetence masquerading as a new art form" and documented several glaring factual inaccuracies.[27]

On the surface, Wolfe's articles are not so bad as the attacks would have us believe. He profusely complimented Lillian Ross, something he could not bring himself to do when he wrote "The New Journalism." Even if intended as parody, some of it reads like the work of a critic—targeting the magazine's habit of publishing long, convoluted sentences, for example. Wolfe called such constructions "the whichy thicket." He criticized writers who imitated

Lillian Ross's style, not important for its own sake, but a fore-grounding of the armies of Tom Wolfe-imitators who would invade journalism in the late 1960s. Toward the end, Wolfe wrote a treatise on the readers of *The New Yorker:* "since the war, the suburbs of America's large cities have been filling up with educated women with large homes and solid hubbies and the taste to . . . *buy expensive things.*"[28] Wolfe said this trend explained a certain bourgeois sentimentality in *New Yorker* fiction: "After all a girl is not really sitting out here in Larchmont waiting for Stanley Kowalski to come by in his ribbed undershirt and rip the Peck and Peck cashmere off her mary poppins."[29]

The antifemale attitude aside, it was formula mid-1960s Tom Wolfe prose, very similar to his attacks on modern art and architecture—but those attacks had not been written yet. Readers could only assume the "Tiny Mummies!" articles were part of his journalistic repertoire, not a piece of cultural criticism voiced from a personal point-of-view.

The crew at *The New Yorker* was livid. Wolfe had reported—even citing Cook County Court records—that William Shawn had been on a list of intended victims prepared by the young killers Leopold and Loeb, and as a result became "retiring." Cook County records do not support this story; Adler and Jonas say it was simply not true. At first, Wolfe defended the pieces. He said in a radio interview that the Leopold and Loeb story was "common dinner-table conversation" although he could not find anything conclusive to confirm it.[30] Later, Wolfe's story changed. In his 1973 essay, Wolfe said the pieces were "lighthearted . . . A very droll *sportif* performance, you understand."[31] This revision of history made little sense, but Wolfe did correctly point out that his critique of *The New Yorker* was *not* New Journalism: "it used neither the reporting techniques nor the literary techniques; underneath a bit of red-flock *Police Gazette* rhetoric, it was a traditional critique, a needle, an attack, an 'essay' of the old school."[32] More accurately, it was a piece of *New Yorker* bashing, such as might be found today in *Spy.* Wolfe has never reprinted the articles in his collections.

At the time, Wolfe's articles generated bitterness that has remained on both sides, and it stimulated several defensive strikes on the New Journalism.[33] Years later, John Hersey and others had still not forgiven him. With the exception of Truman Capote,

New Yorker writers, after the "Tiny Mummies!" episode, did not want their names associated with Wolfe and the New Journalism, which they considered a plague.[34] The episode caused lasting injury. It separated the New Journalism from *The New Yorker* writers. Wolfe and his friends became spokesmen for a form that *The New Yorker* heavyweights would not acknowledge even though they had nurtured it through the postwar years. Without positive criticism and collegial debate, the New Journalism stumbled off in its own direction.

Surprisingly, given all the controversy that "Tiny Mummies!" generated, historians have generally followed Wolfe's interpretation of the birth of New Journalism. The style of writing deserved a name (even if it was not the *first* New Journalism) and Wolfe promoted it. It also deserved a history, which in Wolfe's account was a relatively clean slate—indeed, how can you call something *New* Journalism if its forerunners include several major twentieth century writers and its roots reach to Defoe? Wolfe could not bring himself to acknowledge *The New Yorker*'s genuine contributions.

The magazine created by Ross and Shawn provided the institutional conditions that nourished literary journalism from the late 1930s: time, space, freedom, and financial backing. *The New Yorker* attracted some of the best writers, and gave them an opportunity to create nonfiction literature. This was in a context that differed from the rise of New Journalism. The controversies and generational conflicts of the 1960s did not give *New Yorker* writers an excuse to fling their innovations as a challenge onto their editors' desks. Wolfe may have found writing New Journalism for the magazines to be a mark of status that separated him from newspaper reporters, but Mitchell, Liebling, Ross, and Hersey had more friendly cooperation with newspaper workers. For example, Mitchell said that after his long profile of Mr. Flood apeared, several people went looking for the old fellow in the fish market. When he could not be found, the newspapers said nothing, even though several editors knew Mr. Flood was a composite.

Mitchell and his fellow writers can trace their heritage back through Turgenev, Borrow, and Defoe. Many of them rose through the newspaper feature ranks, as did Wolfe, to become the star nonfiction writers of their time.

Liebling, Ross, Hersey, and especially Mitchell pioneered the styles that a few years later made New Journalism such a notorious and enjoyable literary form. In Mitchell's work on Mr. Flood and Joe Gould, and in his symbolic cultural portraits of lower Manhattan and the surrounding fishing communities, he quietly succeeded in merging fiction and nonfiction, the symbolic and the literal, biography and reportage, the real and the imagined landscapes of the city, in a way that continues to influence and inspire nonfiction writers today.

NOTES

1. Joseph Mitchell's books include *My Ears Are Bent* (New York: Sheridan House, 1938), a collection of his newspaper work; *McSorley's Wonderful Saloon* (New York: Duell, Sloan and Pearce, 1943); *Old Mr. Flood* (New York: Duell, Sloan and Pearce, 1948); *The Bottom of the Harbor* (Boston: Little, Brown and Company, 1960); with Edmund Wilson, *Apologies to the Iroquois With A Study of the Mohawks in High Steel* (New York: Farrar, Straus, 1960), which contains Mitchell's classic ethnographic study of the Indian ironworkers; and *Joe Gould's Secret* (New York: Viking Press, 1965).

2. Raymond Sokolov, *Wayward Reporter: The Life of A. J. Liebling* (New York: Harper & Row, 1980), pp. 104–5.

3. Paul Bush, "The Use of Fiction Elements in Nonfiction: Proving the Existence of a New Genre," Master's degree thesis, Vermont College of Norwich University, 1989, Ch. 3.

4. Mitchell, *My Ears Are Bent*, p. 11.

5. Stanley Edgar Hyman, "The Art of Joseph Mitchell," in *The Critic's Credentials* (New York: Atheneum, 1978), p. 79.

6. Mitchell, *Old Mr. Flood*, p. vii.

7. Malcolm Cowley, "The Grammar of Facts" in *The New Republic* (July 26, 1943), p. 113; also reprinted in *The Flower and the Leaf* (New York: Viking Press, 1985).

8. Unless otherwise noted, Mitchell's remarks were made during several interviews with the author in 1988–1989.

9. Noel Perrin, "Paragon of Reporters: Joseph Mitchell," in *The Sewanee Review,* 91(2):182 (Spring 1983).

10. Hyman, *The Critic's Credentials,* p. 83.

11. Mitchell, *Bottom of the Harbor,* p. 189–90.

12. Mitchell, *Joe Gould's Secret,* p. 68.

13. Noel Perrin, "A Kind of Writing For Which No Name Exists," in *A Reader's Delight* (Hanover, N.H.: University Press of New England, 1988), p. 21.

14. Mitchell, *Joe Gould's Secret,* p. 139.

15. Mitchell, *Joe Gould's Secret,* p. 140–45.

16. Mitchell, *Joe Gould's Secret,* p. 153.

17. Hyman, *The Critic's Credentials,* p. 84–85.

18. Mitchell, *Joe Gould's Secret,* p. 37.

19. Mitchell, *Joe Gould's Secret,* p. 88.

20. Mitchell, *Joe Gould's Secret,* p. 169.

21. Sokolov, *Wayward Reporter,* p. 252.

22. *Normandy Revisited,* copyright 1955, 1956, 1957, 1958 by A. J. Liebling, can be found in *Liebling Abroad* (Playboy Press, 1981).

23. Sokolov, *Wayward Reporter,* p. 10.

24. Personal interview, Sept. 24, 1982.

25. John Hersey, "The Legend on the License," in *The Yale Review* 70:1–25 (Autumn 1980). "Joe Is Home Now" can be found in Hersey's *Here to Stay* (New York: Alfred A. Knopf, 1963), pp. 109–33.

26. Tom Wolfe, "The New Journalism," in *The New Journalism with an Anthology,* Tom Wolfe and E. W. Johnson, eds. (New York: Harper & Row, 1973), pp. 8–9.

27. *Columbia Journalism Review* (Winter 1966), pp. 32–34.

28. Tom Wolfe, "Lost in the Whichy Thicket: *The New Yorker*—II," *New York* Magazine of the *Herald Tribune* (April 18, 1965), p. 24.

29. Wolfe, "Lost in the Whichy Thicket," p. 24.

30. Leonard C. Lewin, "Is Fact Necessary?" in *Columbia Journalism Review* (Winter 1966), pp. 29–34.

31. Wolfe, "The New Journalism", p. 24. Although he was not talking about exactly the same thing, compare his comment in a *Rolling Stone* (Nov.–Dec., 1987, issue 512) interview: "In my mind, I was never satirizing anybody. My intention, my hope, was always to get inside of these people, inside their central nervous systems, and present their experience in print from the inside" (p. 218).

32. Wolfe, "The New Journalism," p. 24.

33. See, for example, the articles in Part Four of Ron Weber's *The Reporter as Artist: A Look at The New Journalism Controversy* (New York: Hastings House, 1974), and John Hersey's "The Legend on the License," from 1980.

34. See especially Hersey, "The Legend on the License," pp. 5–7. Discussing Wolfe's "Tiny Mummies!" articles, Hersey said, "in them one finds in gross form the fundamental defect that has persisted ever since in Wolfe's writing, and that is to be found in the work of many of the 'new journalists,' and also indeed in that of many 'nonfiction novelists'—namely, the notion that mere facts don't matter." This was published in 1980, and is *not* a fair assessment of Wolfe or the New Journalism.

5

The Politics of the New Journalism

John J. Pauly

For more than twenty years journalists, educators, and literary critics have pondered the significance of the New Journalism. Confronted with the flourish and publicity that accompanied the work of Tom Wolfe and his brethren in the mid-1960s, those commentators have searched for New Journalism's soul, for the life-spirit that once burned so brightly, but now seems more or less extinguished. Viewed in retrospect, the New Journalism's lasting significance may have been its ability to gather such commentary, and in so doing condense and intensify vexing conflicts over cultural style. The very term *New Journalism* proved singularly effective at calling out opponents into symbolic combat, for its old–new distinction joined, by chains of implication, several other contested oppositions—adult–child, objective–subjective, reason–emotion, fact–fiction, corporate–personal.[1] More specifically, New Journalism offered a double dare to the establishments of Journalism and Literature. It challenged the authority of Journalism's empire of facts, and the sanctity of Literature's garden of imagination. The New Journalism, as a discourse on the nature of our storytelling practices, confronted both Journalism and Literature with the social habits and institutional structures that sustained them.

That, however, is not exactly the official history of the New Journalism. Following Tom Wolfe's lead, commentators have often restricted the label *New Journalism* to the experimental nonfiction styles produced by him and other writers, such as Joan Didion, Gay Talese, Jimmy Breslin, Norman Mailer, and Hunter Thompson.[2] Unfortunately, that appropriation writes off other meanings that nonfiction writing once held for both writers and readers.[3] It is not so easy to extract the new nonfiction from the turmoil of the 1960s. In the name of the New Journalism, critics once condemned the corporate caution of newswork, reporters' overreliance on official sources, the increasing concentration of media ownership, and biased news coverage of political demonstrations. They proclaimed the need not just for experimental literary styles but for an underground press, advocacy journalism, journalism reviews, and newsroom democracy.[4] Those other issues have now been exiled to the subtropics of scholarly discourse. Even the sense of newness has diminished, for Wolfe and his cohorts have entered respectable middle age. Critics today read nonfiction texts (ever more closely) as instances of a more generic form, called *literary nonfiction* or *literary journalism.*

Whatever happened to the New Journalism? Our critical discourse forged it into a literary canon and, in the process, disarmed its politics. We no longer hear it as a call to arms. Ironically, both friends and enemies of the New Journalism have sanctioned our amnesia. Its enemies now dismiss it as a tie-dyed duffel bag of ornamental doodads, and gleefully heave it overboard along with the other counterculture claptrap of the 1960s. Its friends, hoping to carry it on a while longer, collapse it to fit the more *au courant,* jet-set styles of textual politics. In either case the New Journalism is now deemed excess baggage.

I propose to rearm the New Journalism as a political issue by recovering a fuller memory of it. By politics I do not mean party loyalty, for there is no one banner under which Wolfe, Breslin, Mailer, Didion, and Thompson would ever agree to march (though opposition to the Vietnam War came close). Rather I mean politics as that realm of symbolic confrontation in which groups of citizens organize, enact, and negotiate their relationships with one another. Through activities like writing, publishing, and reading, groups come to imagine one another and thus to constitute the very forms of public life.[5] As a form of cultural politics, the New Journalism persistently disrupted taken-for-granted social relation-

ships between writers, subjects, and readers. That disruption is the
memory we now desperately need to recoup.

Canonization always blesses some remembrances of literary history and excommunicates others. When writers and critics memorialize the new nonfiction as a literary genre, they tend to ignore
the social, political, and economic worlds in which writing (including their own) gets done. At worst they begin to assume that they
can read society off the pages of a text without engaging, except in
silent language, the human actors who inhabit that society. This
tendency to read texts as society (rather than society as a text)
produces a fascination with the New Journalism's narrative qualities. Commentators typically treat each nonfiction story as a discrete work of literary imagination. They examine a work's *style,*
describing its aesthetic or writerly qualities, such as reliance on
generic conventions, or deployment of tropes, point of view, and
description; or they evaluate a work's *truth claims,* noting how the
play of the text purports to create knowledge. Earlier critics read
the New Journalism as a superior representation of reality, a kind
of journalistic deep truth; more recent critics read it as a clever
deconstruction of its own claims to authority. Both approaches
start, and usually end, with a close reading of textual action.

We might also interpret a work of reporting as a *social behavior,*
without precluding close textual analysis. We could study the *venues of publication* (i.e., the institutional sites at which the story was
written, printed, disseminated, and discussed). We could analyze
the research and writing of a work as *social acts,* noting the way the
reporting process implicates writer, subjects, and readers in relationships beyond the text. We could also interpret *commentaries*
that make a work into a sign of the times, and attribute wider
meanings to the acts of producing or consuming it. Social analysis
does not replace the text itself, but amplifies a text's play of meaning by "reading" writing as a form of work, performed under particular conditions.

Critics' concentration on literary technique transforms the history of the New Journalism into a triumph of individual sensibility.
Landmark works chart the new genre's path—Talese's profile of
Joe Louis, Truman Capote's *In Cold Blood,* Wolfe's *The Kandy-
Kolored Tangerine-Flake Streamline Baby,* and Mailer's *The Armies of the Night.* Even discussions that allude to the social context
of nonfiction do so only to stress the significance of individual

achievement. In such accounts several prominent writers (*prominent* being variously defined as most successful, hip, popular, influential, stylish, or avant garde) come to represent "the age." The book-length studies of the new nonfiction, for example, often adopt this approach. John Hollowell chooses Capote, Mailer, and Wolfe for his particular chapter studies.[6] John Hellmann chooses Mailer, Thompson, Wolfe, and Michael Herr.[7] Chris Anderson chooses Wolfe, Capote, Mailer, and Didion.[8] Everette Dennis and William Rivers choose Wolfe, Lillian Ross, Talese, Breslin, Capote, and Mailer as representative practitioners of their "new nonfiction."[9] Even commentators who probe the cultural context of the New Journalism borrow this method. Morris Dickstein's essay on "The Working Press, the Literary Culture, and the New Journalism" begins with a sociology of contemporary writing, but ends with a long-winded discussion of Tom Wolfe and Norman Mailer as polar ideal types.[10]

On one hand this procedure makes perfect sense. After all, those eight or ten writers are the ones whose work attracted the greatest attention (or criticism). Furthermore, in terms of talent and technique, those writers are arguably among the most skillful practitioners of their era. On the other hand, this vest-pocket memory of the New Journalism may be tailored to other needs as well. For the popular media, this Great-Men-of-the-Decade method creates a sense of breathless social change (history as a quick read). For academic critics, the method efficiently meets occupational demands for productivity (history as the stuff of tenure).[11] For citizens this method of cultural synecdoche provides institutionally packaged remembrances, but not a shared memory (history as a dead letter).

The critical discourse on the New Journalism thus produces a canon first by stressing the achievements of a few good men (and one or two women). Next critics affirm those choices by identifying the handful of essential storytelling techniques the great works display. Again no one could deny that some criticism of technique is essential. Writers experience their work, at least in part, as the resolution of technical problems through choice of point of view, structure, voice, and description. But the fascination with technique, like the search for Great Artists, also serves other social purposes. It helps rescue the reputation of the New Journalism (and its supporters) by making that writing more "literary" (con-

versely, detractors can use technique to condemn nonfiction reporting for not being anything new at all). The search for the New
Journalism's literary essence, however, cannot discover much
about what and how styles of writing meant. Throughout the
1960s, journalistic style was contested political terrain, and compet-
ing groups fought to capture or defend it. Critics of the New Journalism who wished to deny everything it stood for often chose style
as their point of attack. If the New Journalism was not really new,
they argued, young writers' claims could be dismissed as merely
the boasts of brash, ignorant youth.[12] Such arguments disappear
from memory when we memorialize the New Journalism as a sequence of discrete aesthetic experiments, loosely held together by
"the times."

One important example of how literary style signified social contention was the protracted early debate over the New Journalism's
"personalism." Journalists who wrote in a distinctive personal
voice wanted to be free to tell stories as they saw them, without
being shackled by institutional conventions of objectivity. They
thought that personal involvement and immersion were indispensable to an authentic, full-blooded account of experience. Detractors, especially editors, abhorred that personal voice, because they
felt it betrayed the public's trust in journalism as unbiased fact.
The controversy clearly went beyond narrow matters of literary
technique, for it alluded to ubiquitous arguments about journalists' sincerity, authenticity, and loyalties. "Personalism" articulated
young journalists' newly imagined connection to the people and
events they reported. As a style of political commitment, personalism opposed itself to the thin-lipped, emotionally repressive
style of middle-class worklife. In journalism that conflict over cultural style often played itself out in newsroom confrontations that
were at once generational as well as professional.[13] Young reporters refused to remain laptop dummies who lip synched the newspaper's institutional voice. They demanded official recognition that
they were the ones in the streets, close to the real action. Editors
resisted such nonnegotiable demands. To them personalism was
the chant of a permissive and disrespectable age, in which lazy,
self-indulgent young reporters refused to accept their professional,
institutional responsibilities.[14]

The New Journalism's versions of truth often aroused similar controversy. Alarmingly to its detractors, delightfully to its defenders, the New Journalism played along the border of fact and fiction. Truman Capote's *In Cold Blood* elicited obsessive attempts to expose each instance of his fictionalizing.[15] Tom Wolfe's wildly satirical and footloose story about *The New Yorker* brought forth venomous responses by Dwight MacDonald, Renata Adler, and other New York literati.[16] Gail Sheehy's admission that her prostitute "Redpants" was a composite character drew wide criticism.[17] Janet Malcolm's recent expose on Joe McGinniss's interviewing ploys enacts this ritual once more.[18] For a time, each incident has constituted a loyalty test to distinguish the friends of the New Journalism from its enemies. John Hersey, in a relentlessly high-minded attack on the young whippersnappers of New Journalism, summarized the outrage of his generation. He exhorted writers to note the social purposes for which they had been "licensed." Hersey gravely reminded everyone that the legend on the journalist's license said "*None* of this was made up."[19]

Later I will discuss more fully the nature of journalistic truth. For now I will only note that journalism's crisis of authority echoed similar crises throughout American society. The fact–fiction debate is not fully intelligible without reference to the New Journalism's political moment. In that volcanic age, complaints about unfairness in representation rained down on all sides. United States government policy in Vietnam introduced the term *credibility gap,* and protracted family conflicts added the term *generation gap* to the popular vocabulary. As society's self-appointed symbol-broker, journalism found itself at the epicenter of such debates. Spiro Agnew, in his famous 1969 speech in Des Moines, stepped up the Nixon administration's systematic attack on "the media." Students protested the press's coverage of political demonstrations.[20] Professional journalism's truth, which never stands unchallenged, has rarely been as fiercely contested as it was in the late 1960s and early 1970s.

A 1971 cartoon in the American Society of Newspaper Editors' *Bulletin* dramatized journalists' sense of their own predicament. That cartoon portrayed a panicky white male newpaper editor as Gulliver, tied down under a front page headlined "More Bad News," and being harassed by a headbanded hippy, a braless woman demanding "Women's Lib," an afro-ed black demanding

"Black Power," Indians demanding that society "Give Back the U.S.A.," a masked, pocket-picking criminal, an angry American soldier, an inscrutable Vietnamese, and a war protestor who, having stuck an "End the War" sign in the editor's ear, was now giving him the raspberries.[21] That cartoon reminds us that though we often argue the truth of texts on the high ground of philosophy in the comfort of graduate seminars, writers and readers in the late 1960s fought for truth at the family dinner table and the union hall, and in newsrooms, boardrooms, barrooms, and the streets.

So far I have argued that we miss powerful meanings of the New Journalism if we remember it only as a literary canon, and that a critical reading of selected writers' works cannot capture those meanings without a parallel exegesis of the New Journalism as a "sign of the times." I now wish to outline a reinterpretation of the New Journalism as a social rather than an aesthetic discourse. My account of the New Journalism as a politics of cultural style will stress its character as a *social act* that was conducted in particular *venues* and that elicited *commentaries* that symbolically positioned opposing groups.

During the late 1960s and early 1970s, the New Journalism spoke for a social movement that aimed to transform not just the styles of nonfiction writing, but the very institutions through which society produced and consumed stories about itself. For friend and foe alike, the new nonfiction's revolt into style often carried political meanings. Just as a generation of Muckrakers had promoted the politics of managerial professionalism during the Progressive Era, so did New Journalists promote the politics of youth. Indeed, the most obvious link between the new nonfiction and other journalistic experiments, such as underground papers, newsroom democracy, and media reviews, was the youthfulness of the reformers. The New Journalism was a literary movement rooted in generational conflict. The two lists that follow demonstrate my point. Table 5.1 lists the birthdates of all the writers in Wolfe's 1973 anthology and their ages in 1965, the year often proposed as the birthdate of the New Journalism. Table 5.2 lists similar information for a group of nonfiction writers, many of whom worked for the *The New Yorker* or the *The New York Times,* and whose work various commentators thought was part of the same general cultural mood.

TABLE 5.1. Birth Years of Writers in Wolfe's Anthology

Name	Year of Birth	Age in 1965
Truman Capote	1924	41
Robert Christgau	1942	23
Joan Didion	1934	31
John Gregory Dunne	1932	33
Joe Eszterhas	1944	21
Barbara Goldsmith	1931	34
Richard Goldstein	1944	21
Michael Herr	1940?	25
Norman Mailer	1923	42
Joe McGinniss	1942	23
James Mills	1927	38
George Plimpton	1927	38
Rex Reed	1938	27
John Sack	1930	35
"Adam Smith"	1930	35
Terry Southern	1926	39
Gay Talese	1932	33
Hunter Thompson	1939	26
Garry Wills	1934	31
Tom Wolfe	1931	34

The proportion of writers in their mid-twenties to mid-thirties is remarkable. The tables suggest that a whole generation of nonfiction writers came of age in the mid-1960s. Those we now call New Journalists generally achieved their fame through their work for magazines such as *Esquire, New York, Harper's, Rolling Stone,* and *The New Yorker.* A large cohort was also working at major newspapers, especially at *The New York Times,* and they have subsequently moved into book-length nonfiction and syndicated newspaper commentary.

Such a listing need not imply some mysterious generational force or historical impulse that drove all these writers into nonfiction. The lists simply represent commentators' sense of the state of the art in nonfiction. The widespread perception that the New Journal-

TABLE 5.2. Birth Years of Other "Representative" Nonfiction Writers

Name	Year of Birth	Age in 1965
Jimmy Breslin	1930	35
Frances FitzGerald	1940	25
Marshall Frady	1940	25
David Halberstam	1934	31
Pete Hamill	1935	30
Larry L. King	1929	36
J. Anthony Lukas	1933	32
John McPhee	1931	34
Willie Morris	1934	31
Jack Newfield	1939	26
James Ridgeway	1936	29
Dick Schaap	1934	31
Jonathan Schell	1943	22
Neil Sheehan	1936	29
Robert Sherrill	1925	40
Gloria Steinem	1934	31
Nicholas von Hoffman	1929	36
Dan Wakefield	1932	33
Tom Wicker	1926	39

ists were young writers inspired younger journalism students in universities, who often found little appealing in the occupational identities or work that conventional journalism offered.[22] For their part newspaper editors returned this disdain, complaining about the young people who were going into newswork. Young journalists, they said, expected too much too soon. They were undisciplined and unkempt people who showed up for work with long hair, wore lapel buttons advocating controversial causes, and demanded that the newspaper right social wrongs. The more gentle souls among the editors sometimes counseled patience with the young, and the braver ones even noted the hypocrisy of the establishment's own ethics, which encouraged editors and publishers to support the United Way and the Chamber of Commerce but not SNCC or Mobe. In the end, however, editors treated young reporters as a managerial problem, to be handled either with old-fashioned discipline or heart-to-heart "rap" sessions.[23]

In that context writing and reading *new* journalism affirmed a generational identity. For example, Wolfe's famous introduction to his anthology can plausibly be read as a generational manifesto. For Wolfe the New Journalism signified socially as well as aesthetically. By tracing his geneology to the newspaper feature writers of the 1950s, Wolfe deliberately aggravated news journalists by devaluing the investigative reporting or political commentary to which they had staked their identities. Reading the New Journalism also articulated a cultural identity. The new nonfiction seemed hip and relevant in ways that the daily newspaper had long since ceased to be. For readers who aspired to be writers, the New Journalism symbolized an economically as well as aesthetically liberated style of work. Released from the endless tedium of writing for special interest magazines, the free-lance writer would now roam the country in search of the American zeitgeist (and get paid handsomely in the process).[24] This romantic vision of the writer, which the New Journalism rendered plausible, annoyed workaday journalists, who found it absurd and pretentious, perhaps because they had long since abandoned such hopes themselves.

The politics of youth alone could not have produced the New Journalism, however, without the simultaneous development of venues in which to publish experimental styles of reporting. The new nonfiction capitalized on a general trend in the magazine business—a turn from fiction to nonfiction in the years after World War II. The small number of publications that regularly sponsored major works of New Journalism, or the exceptional circumstances that sometimes led to such experiments, are rarely noted. For example, Sharon Bass and Joseph Rebello have demonstrated that *Esquire* magazine's turn to nonfiction in the early 1960s was part of a desperate editorial response to strong competition from *Playboy*.[25] The failure of *Esquire*'s older formula opened up a space in which editors were willing to risk unusual styles of reporting. In his four-year regime at *Harper's*, Willie Morris accomplished a similar feat.[26] The death of the traditional short story market meant that more professional "literary" writers were willing to try nonfiction.

Similarly idiosyncratic circumstances attended the birth of the New Journalism in its other major venues. *New York* magazine started as the Sunday supplement of a dying New York *Herald Tribune,* and survived as the pioneer of a new magazine genre. *The Village Voice,* founded in 1956, prefigured later upscale urban weeklies; it survived by catering to an intensely self-conscious, compact intellectual-artistic community that had no equivalent anywhere else in the United States. *Rolling Stone,* born in 1968, discovered a lucrative new source of advertising support in the emerging rock'n'roll culture. *The New Yorker* magazine, long sustained as the genteel voice of Manhattan culture by its extravagant advertising base, continued to bankroll long reporting projects. Such magazine work undoubtedly appealed to many writers because it lent itself more readily than newspaper writing did to republication in book form, and the success of such books freed at least a small number of prominent writers to go it alone.

One other economic aspect of the New Journalism deserves further study. From the start, the new nonfiction was predominately a New York phenomenon (as indeed the entire history of literary nonfiction has been, to some degree). In effect, the canonization of New Journalism tells once more the story of New York as the hub of the national culture industry and Mecca for ambitious writers. Markets elsewhere in the country have rarely been able to afford the financial extravagance of immersion reporting. The underground press and political papers like *Ramparts* sometimes copied the style of the New York nonfiction but could not match its expensive habits of research and fact-checking. Small papers generally find it impossible to support professional writers of any sort for very long.[27] Thus the market for New Journalism has displayed a predictable symmetry: a few major magazines, mostly in New York, willing and able to do in-depth nonfiction; many small magazines willing but unable to finance it; a few major newspapers able to finance it, but unwilling to do so; and many small newspapers neither able nor willing to support it.

In today's New York market, even fewer sponsors can afford immersion reporting. *The New Yorker*'s advertising base has long sustained its reporters, although it may not for much longer. *Esquire* and *Harper*'s have sporadically funded such work. The highly specialized audiences of the *Village Voice* and *Rolling Stone* continue to sustain them, although the thrill is gone from both of those publications. Much stylish nonfiction writing has moved into the city and regional magazines, where it has often had to check its politics at the door. Many of the original New Journalists, having made their individual reputations, move easily now from one book to the next, living off royalties, publishers' advances, and speakers' fees. In this sense these prominent New Journalists have followed the same career path cleared by previous generations of nonfiction writers. Magazine writers use their articles to capture a publisher's attention and win lucrative book contracts. In turn, the publishers use magazine articles to gauge the potential marketability of a writer's work. New Journalism in its original form—in-depth nonfiction reporting for magazines—increasingly looks like a handmade good in an age of mechanical reproduction, an expensive taste that only a few prominent publications can indulge.[28]

In short, the social, economic, and political moment of the New Journalism has disappeared, leaving only traces of its existence—a canon of official works, marketable reputations for a few writers, on the left a nostalgia for the recklessness of lost youth, on the right a chorus of smug I-told-you-sos. Amid the rubble of the New Journalism, can we find the materials with which to build a usable past? In fact, we will close the book on the politics of the New Journalism if we treat it as a canon of individual accomplishments. But we could remember the New Journalism as something else—a discourse about nonfiction writing as a social practice. Its contradictory example could continue to warn us of the possibilities and problems of reconstructing research and writing as truly democratic activities.

Part of the New Journalism's distinctive political contribution was certainly its fascination with culture.[29] In the traditional newsroom, culture had always remained a residual category, the social stuff to be cleaned up by female writers after real men had devoured the political issues of the day. In the New Journalism, however, culture—often experienced as the politics of style—supplied the very substance of the reporting, and the attempt to report on culture usefully complicated discussions about the truth of nonfiction writing. After the New Journalism, reporters could no longer appeal unselfconsciously to "facts" to justify their stories.[30] Journalism's fact-fetish, we now understand, serves in part as a political strategy for displacing community complaints about reporters who, by the very nature of their occupation, mediate groups' understandings of one another. Reporters use facts rhetorically, not just to defend their interpretations against attack, but to naturalize and objectify those interpretations.[31] Readers or sources who dispute those interpretations turn the same strategy back on reporters, condemning the story because it "got the facts wrong," thereby implying that the truth was there all along if only the reporter had chosen to look for it. In debates over the news, each group uses "reality" to bludgeon the other's imagination.

Literary critics enjoy debunking the realism of nonfiction stories, for they hope to affirm the fictiveness of all narratives. Having settled journalism's hash, philosophically speaking, critics can deny all claims to representation, and hence free the literary imagination from its earthly entrapments. I would agree that all narratives are fictions, and that realism mostly means a set of shared stylistic conventions for dramatizing authenticity. I would also maintain that the New Journalism offered something as a form of journalism, not just as a disguised, inferior form of fiction. The New Journalism can still remind us that the truth of all writing is a matter for social negotiation. To say a report is true is to affirm that it speaks the consensus of some actual community of interpreters, who read the social conditions in which the story was produced as well as its narrative strategies. In turn, disagreements over truth signal appeals to different communities of interpretation, with their own standards of evidence, significance, and style.[32]

Thus the strongest objection to conventional journalism's truth is not that it claims to be representational or objective, but that it unilaterally asserts rather than fully argues its truth claims. The strength of a newspaper's interpretations are rarely tested by a community of readers. Because the daily newspaper carries an astonishing number and variety of stories each day, single stories escape the scrutiny that has been directed toward many New Journalism stories. Individual readers have neither the time nor the knowledge to verify the truthfulness of all the stories each day, nor are the newspaper's consumer-readers organized into a stable oppositional community that could provide the psychic, informational, or financial resources to dispute the accounts of large news organizations. Newspaper editors know from their own experience that the most skeptical readers are those who have witnessed an event and can compare their personal experience of it to the news coverage. The letters-to-the-editor column probably only hints at readers' disagreements with their daily newspapers.

The truth of journalism does not reside in its representationalist narratives, as journalists and literary critics both assume. Writers use conventional codes to convey truth, but such codes are themselves just one form of a larger series of social occasions during which interpreter and interpreted meet to argue their positions. Literary critics typically reinvent the relationships of writer and reader in each text, without much continuity from one text to another, and without reference to the nonliterary social worlds that writer, readers, sources, and subjects cohabit. In effect, literary criticism's thin description of writer–reader relationships reproduces the social structure of the capitalist marketplace, in which writers know readers only as an abstracted market.[33] Contemporary criticism's speculations about reader response, implied readers, oppositional decodings, and the rest, are not entirely satisfactory attempts to describe relationships between writers and readers in a world where the two need never meet. The political turmoil of the 1960s, however, made it impossible for journalists, at least, to maintain the social distance that modern writers enjoy and so assiduously cultivate.[34] As Walter Ong reminds us, reasonably enough, the reader of a text is always a fiction.[35] Yet the readers of nonfiction are actual creatures, too (so are fiction's readers, however inconvenient their existence may prove to writers). Actual people who are written about do not always sympathize with the writer's claim that the world exists to provide grist for his mill.

This then was the crucial political contribution of the New Journalism: for a time it compelled us to reconsider the social rituals through which writers might declare themselves responsible rather than merely free. Early defenses of the New Journalism repeatedly emphasized reporters' need to answer to their subjects and readers in ways that newspapers could not (or would not). The New Journalism invented a variety of strategies for mediating writer's relationships to everyday life, actual readers, and subjects. Some of those strategies, such as submitting the draft of a complex story to a source for fact-checking, involved modest but important changes in procedure. In general the prominent nonfiction that we now call the New Journalism actually worked in a social netherworld between the elegant abstractions of the conventional press and the tribal loyalties of the alternative press. Conventional journalism treated writer–reader relationships as a credibility problem, to be resolved by appeals to higher cosmopolitan ideals such as objectivity. The alternative press imagined a wholly different covenant, better described as *communitarian,* that connected it to readers living in specific local worlds. This tension between cosmopolitan and communitarian occupational styles—the conflict between work done for the marketplace and in the subculture—produced the New Journalism's distinctive style of cultural politics.[36]

By this path the argument about the New Journalism returns to a discussion of narrative, only with a better sense of the world. The narrative strategies of the New Journalism now appear not as instances of universal "discourses," but as styles of social action, traces of the life-worlds in which particular writers finesse their relationships with the marketplace, sources, and readers. The economic production of the prominent works of New Journalism remained predominately cosmopolitan; professional writers working out of New York City created elaborate stories for nationally distributed commercial magazines. The New Journalism also spoke the concerns of oppositional groups in a tone rarely heard in the daily press.

In the end, the New Journalism's attempt to balance cosmopolitan and communitarian styles of journalism proved unstable. Everette Dennis, an early chronicler of the New Journalism, spoke the hopes of a generation when he dreamed of a journalism that would be reportorially stylish as well as politically astute.[37] But that recurring political dream of the left, in which avant-garde art allies itself with grassroots politics, soon disappeared once more. For the most part the new nonfiction created a style of cultural politics that simulated rather than accomplished participation. Consider the long and distinguished record of *The New Yorker,* a magazine that has steadfastly sponsored fine reporting, sometimes even at economic cost to itself.[38] In effect, that magazine's insistence on meticulous factual accuracy imagines an ideal reader to whom the magazine feels responsible. For a magazine with a national constituency, that reader may be a useful, even an indispensable, fiction. In most respects, however, *The New Yorker* offers a quite conventional model of journalism and of the community of interpretation. Its legendary fact-checking procedures invest the authority of the text in the editor rather than the reporter. The magazine assumes, with some justification, that its readers share its standards of meticulousness. *The New Yorker*'s typical method—assigning a professional writer to "do" part of the culture in a comprehensive story or series—enacts a cosmopolitan style of knowingness that flatters readers.

That is one approach to journalism, but it is not the only approach, nor even a universally admirable ideal. An alternative form of journalism, for example, might imagine a more conversational relationship between writers and readers. Readers would inevitably continue to chide writers for careless handling of details, not because stories fall short of some factual ideal, but because such carelessness bespeaks indifference to or disrespect for the actual groups being written about. In such publications readers as well as writers might benefit from a less intrusive editorial presence. A publication could simply decide to stand back a bit while writers and readers provoke, dispute, resolve, and celebrate their misunderstandings of one another. Most editors, reporters, and publishers would deny the name *journalism* to such an alternative publication. For them writing must remain a full-time, cosmopolitan profession, conducted on citizens' behalf by commercially profitable publications that bureaucratically organize the symbolic practices through which a community represents itself.

The New Journalism never displaced conventional styles of journalism; it probably never intended to, even in the fevered imagination of Tom Wolfe. For a time, the New Journalism did compel a discussion of the ways in which modern groups encounter one another through writing, publishing, and reading. As a style of cultural politics, the New Journalism forced journalists and fictionalists alike to confront what it means to be a writer and to be written about, what writers owe their subjects and readers, and by what habits society organizes its practices of public imagination.[39] That this memory of the New Journalism has been lost in transit, packaged and hurriedly shipped off to Literary Historyland, ought to remind us of how difficult was the revolution we once hoped to win.

NOTES

1. David Eason has demonstrated the subtle persistence of such oppositions in discussions of the Janet Cooke case. See "On Journalistic Authority: The Janet

Cooke Scandal," *Critical Studies in Mass Communication* 3:429–47 (December 1986); reprinted in *Media, Myths, and Narratives,* James W. Carey, ed. (Newbury Park, Calif.: Sage Publications, 1988), pp. 205–27.

2. Even Tom Wolfe apparently debated whether to title his anthology *The New Journalism.* In the biographical note that accompanied an earlier version of his introductory, Wolfe is said to be working on a book called "The New Nonfiction." Tom Wolfe, "The New Journalism," *The Bulletin of the American Society of Newspaper Editors,* No. 544 (September 1970), p. 18. Several years later, in a talk on contemporary literary technique at the University of Michigan, Wolfe avoided the term altogether. He ends by saying, "Ah, but this brings me dangerously close to a topic which I swore five years ago I would never publicly expound upon again . . . the New Something-or-other . . . and so I will stop now." Tom Wolfe, "Literary Technique in the Last Quarter of the Twentieth Century," *The Michigan Quarterly* 17:463–72 (Fall 1978).

3. Wolfe does describe the New Journalism as a style of occupational identity that occasioned status battles among writers. That explanation, true as it is, resists any meanings of the New Journalism to those outside of the New York writing community.

4. The range of meanings attached to the term *New Journalism* is evident in early treatments of the movement. See, for example, Michael L. Johnson, *The New Journalism* (Lawrence, Kansas: University of Kansas Press, 1971); Everette E. Dennis and William L. Rivers, *Other Voices: The New Journalism in America* (San Francisco: Canfield Press, 1974); Charles C. Flippen, ed., *Liberating the Media: The New Journalism* (Washington, D.C.: Acropolis Books Ltd., 1974); and James Ridgeway, "The New Journalism," *American Libraries* 2:585–92 (June 1971).

5. For a fine recent study of politics as a symbolically constituted realm, see Murray Edelman, *The Political Spectacle* (Chicago: University of Chicago Press, 1988).

6. *Fact and Fiction* (Chapel Hill: University of North Carolina Press, 1977).

7. *Fables of Fact* (Urbana, Ill.: University of Illinois Press, 1981).

8. *Style as Argument* (Carbondale, Ill.: Southern Illinois University Press, 1987).

9. Dennis and Rivers, *Other Voices.*

10. *The Georgia Review* 30:855–77 (Winter 1976).

11. Charles Newman has discussed the inflated expectations and claims of contemporary literary theory in *The Postmodern Aura* (Evanston, Ill.: Northwestern University Press, 1985). One simple example of Newman's point—critics often title their arcane commentaries in astonishingly broad and provocative ways. No truth-in-packaging law applies to scholarly research, however, as readers cynically come to understand. Pushed by university demands and pulled by personal ambition, professors inflate the significance of their work, but accomplish that work rapidly and parsimoniously.

12. For examples of the debate over the New Journalism's newness, see George A. Hough III, "How 'New'?" *Journal of Popular Culture* 9:114–21 (Summer 1975); Jay Jensen, "The New Journalism in Historical Perspective," in *Liberating the Media,* pp. 18–28; and the following selections from *The Reporter as Artist,* Ronald Weber, ed. (New York: Hastings House, 1974): Lester Markel, "So What's New?,"

pp. 255–59, Michael Arlen, "Notes on the New Journalism," pp. 244–54, and Jack Newfield, "Is There a 'New Journalism'?" pp. 299–304.

13. In symbolic battles over occupational identity, Hunter Thompson has often played an interesting role. Even to very conventional reporters, Thompson's wildly comic, kick-out-the-jams, let-the-truth-be-heard style has made him a folk hero. Similarly Thompson is an editor's worst nightmare—an unreliable, disrespectful misfit and pretender. Thompson himself enacts both personas in his stories, which can be usefully read as an extended meditation on the meaning of being a "professional." For discussions of Thompson's image, see Timothy Crouse, *The Boys on the Bus* (New York: Random House, 1973), pp. 311–14, and Joseph Nocera, "How Hunter Thompson Killed New Journalism," *Washington Monthly* 13:44–50 (April 1981).

14. The most acute discussion of personalism remains Dan Wakefield, "The Personal Voice and the Impersonal Eye," in Weber, ed., *The Reporter as Artist*, pp. 39–48. Also see, in Weber, Nat Hentoff, "Behold the New Journalism—It's Coming After You," pp. 49–53, and Herbert Gold, "On Epidemic First Personism," pp. 283–87; and I. William Hill, "Subjective Jottings on Objectivity," *The Bulletin of the American Society of Newspaper Editors*, No. 537 (January 1970), pp. 8–9.

15. Phillip K. Tompkins has conducted the most compulsive counter investigation. See "In Cold Fact," in *Truman Capote's In Cold Blood: A Critical Handbook*, Irving Malin, ed. (Belmont, Calif.: Wadsworth Publishing, 1968), pp. 44–58.

16. Adler and Gerald Jonas's response to Wolfe was printed in Leonard C. Lewin, "Is Fact Necessary?" *Columbia Journalism Review* 4:29–34 (Winter 1966). MacDonald's response appeared as "Parajournalism II: Wolfe and The New Yorker," *New York Review of Books* 3:18–24 (February 3, 1966).

17. "The Hooker's Boswell," *Newsweek* 80:61 (December 4, 1972), discusses Sheehy's techniques of reporting.

18. "The Journalist and the Murderer," *The New Yorker* (March 13, 1989), pp. 38–73; (March 20, 1989), pp. 49–82.

19. *The Yale Review* 70:1–25 (Autumn 1980). Hersey's sense of indignation is stronger than his memory. Tom Connery has pointed out to me that in "Joe Is Home Now" and "A Short Talk with Erlanger," Hersey admits and justifies his own use of composite characters: "Something needs to be said about the reportorial technique used in this story and the one that follows. These two accounts, unlike the orthodox journalistic tales that constitute the rest of the book, are dovetailings, in each case, of the actual experiences of a number of men"(107). And on the story of Joe, "The story of this veteran's struggles is not 'fictionalized,' because nothing was invented; it is a report. Joe does and says things that were actually said and done by various of the men with whom I talked; I simply arranged the materials"(108). Hersey defends his use of composites as a way to protect the identity of actual returning veterans—a worthy motive, but one that seemingly would require an asterisk on his license. Both stories are reprinted in *Here to Stay* (New York: Alfred A. Knopf, 1963). Norman Sims has also called attention to Hersey's contradictory defense of composite characters in "Origins of the New Journalism: *The New Yorker* Nonfiction Writers," a paper presented at American Journalism Historians Association convention, St. Paul, Minnesota, September 30 to October 4, 1987, p. 10.

20. For examples, see Todd Gitlin, *The Whole World Is Watching*, (Berkeley:

University of California Press, 1980), pp. 40–60, and John Breen, "Bedlam on Campus?" in *Our Troubled Press,* Alfred Balk and James Boylan, eds. (Boston: Little, Brown and Company, 1971), pp. 55–58.

21. *The Bulletin of the American Society of Newspaper Editors,* No. 550 (April 1971), pp. 4–5.

22. See, for example, Roger Rapoport, "A Young Journalist Puts His Typewriter Where His Mouth Is," *The Bulletin of the American Society of Newspaper Editors,* No. 552 (July–August 1971), pp. 3–5; Steven Levine, "Are Newspapers Turning Young People Off?" *The Bulletin of the American Society of Newspaper Editors,* No. 538 (February 1970), pp. 7–9; and David McHam, ". . . Old Ain't Necessarily Good, Either!" *The Bulletin of the American Society of Newspaper Editors,* No. 556 (January 1972), pp. 3–6.

23. For a convenient summary of editors' problems in handling the new generation of reporters, see Norman A. Cherniss, "How to Handle the New Breed of Activist Reporters," *The Bulletin of the American Society of Newspaper Editors,* No. 541 (May 1970), pp. 1–4.

24. André Fontaine, *The Art of Writing Nonfiction* (New York, Thomas Y. Crowell Co., 1974) honestly deals with the problems of free-lancers, but nonetheless exudes enthusiasm about the possibilities for free-lancers who do interpretive journalism.

25. S. M. W. Bass and Joseph Rebello, "The Appearance of New Journalism in the Sixties *Esquire:* A Look at the Editorial Marketplace," a paper presented at American Journalism Historians Association convention, St. Paul, Minnesota, September 30 to October 4, 1987.

26. For accounts of the rise and fall of Willie Morris, see Stuart Little, "What Happened at *Harper's,*" *Saturday Review* 54:43–47, 56, (April 10, 1971), and Charlotte Curtis's article for *[more],* "An Adventure in 'the Big Cave,' " reprinted in *Stop the Presses, I Want to Get Off!,* Richard Pollak, ed. (New York: Random House, 1975), pp. 243–58.

27. Papers such as the weekly *In These Times* survive today only because of heavy voluntary subsidies by readers.

28. In *The Curse of Lono* (New York: Bantam, 1983), Hunter Thompson predicts financial hard times for the independent journalist: "Journalism is a Ticket to Ride, to get personally involved in the same news other people watch on TV, which is nice, but it won't pay the rent, and people who can't pay their rent in the Eighties are going to be in trouble. We are into a very nasty decade, a brutal Darwinian crunch that will not be a happy time for free-lancers"(57).

29. The New Journalism's fondness for popular culture made it the target for a barrage of eighteen articles edited by Marshall Fishwick and published in *The Journal of Popular Culture* 9:99–249 (Summer 1975).

30. For useful summaries of the case against facts, see E. L. Doctorow, "False Documents," *American Review* 26:215–32, (November 1977), and Hayden White, "The Fictions of Factual Representation," in *The Literature of Fact,* Angus Fletcher, ed. (New York: Columbia University Press, 1976), pp. 21–44.

31. Theodore L. Glasser and James S. Ettema, "Investigative Journalism and the Moral Order," *Critical Studies in Mass Communication* 6:1–20 (March 1989), dis-

cusses the process by which investigative reporters use facts to objectify the moral judgments their stories inevitably require.

32. Drawing on speech act theory, Eric Heyne draws a useful distinction between two kinds of truth: a text's *factual status*—whether the writer intends a text to be taken as factual—and its *factual accuracy*—whether readers take that text to be reliable for their purposes. "Toward a Theory of Literary Nonfiction," *Modern Fiction Studies* 33:479–90 (Autumn 1987). A fictional text has neither factual status nor accuracy, Heyne says. A nonfiction text has a factual status, but its factual accuracy is a topic for public debate. Heyne's distinction accomplishes two useful purposes: (1) it avoids an essentialist definition of literature, which grants or denies literary merit to nonfiction texts, depending on the nature of their narrative strategies alone, and (2) it recovers questions of truth as matters of *social* dispute, thus avoiding both a specious fact–fiction distinction and a politically paralyzing reduction of all stories to fiction.

33. The term *contracts* is the popular metaphor for writer–reader relationships in much contemporary literary criticism, including that written about the New Journalism. See, for example, Hellmann, *Fables of Fact,* pp. 1–20.

34. Doctorow admits that "as a writer of fiction I can see the advantages to my craft of not having a reader question me and ask if what I've written is true—that is, if it really happened." "False Documents," p. 218. Like many fiction writers, Doctorow justifies the social distance between fiction writer and reader with a romantic reading of the political subversiveness of fiction. Fact, Doctorow says, is "the power of the regime"; fiction is "the power of freedom"(216). That sort of self-serving nonsense—in which commercially prosperous authors pretend that they are enemies of the State—speaks as much about the social isolation of modern writers as it does about the allegedly transcendent virtues of fiction as a medium.

35. "The Writer's Audience Is Always a Fiction," in *Interfaces of the Word* (Ithaca, N.Y.: Cornell University Press, 1977).

36. My discussion of cosmopolitan versus communitarian occupational styles draws on Douglas Birkhead's discussion of Howard Ziff's distinction between cosmopolitan and provincial styles of journalistic performance and ethics. See Birkhead, "An Ethics of Vision for Journalism," *Critical Studies in Mass Communication* 6:283–94 (December 1989). Also see Ziff, "Practicing Responsible Journalism: Cosmopolitan vs. Provincial Models," in *Responsible Journalism,* Deni Elliott, ed. (Newbury Park, Calif.: Sage, 1986), 151–66.

37. "Journalistic Primitivism," *Journal of Popular Culture* 9:122–34 (Summer 1975).

38. See Ben Bagdikian, *The Media Monopoly* (Boston: Beacon Press, 1987), pp. 105–13, for a discussion of *The New Yorker*'s dramatic loss of advertising revenues in the wake of Jonathan Schell's critical reports on the Vietnam War in summer 1967.

39. None of my comments should be taken as a defense of censorship. To my mind writers' behaving irresponsibly toward the groups they write about is not grounds for revoking their freedom, only for reassessing the value of their contribution.

PART II

6

The Borderlands of Culture: Writing by W. E. B. Du Bois, James Agee, Tillie Olsen, and Gloria Anzaldúa

Shelley Fisher Fishkin

While conflicts over physical territory are usually resolved by force or by negotiated treaty, few comparable mechanisms have been devised for resolving conflicts over cultural territory—of choosing whose realities become reified and whose get "redlined." He who wields political and economic power usually manages to control the power to name and narrate as well. A vast substrata of social, cultural, economic, and historical conditions make it seem "natural" that the dominant paradigms in the culture should dominate.

W. E. B. Du Bois, James Agee, Tillie Olsen, and Gloria Anzaldúa were all concerned with stories that did not fit the dominant paradigms. They were stories of people who were dismissed and devalued because they had the "wrong" race, class, gender, ethnicity, or sexual preference. They were stories of the powerless, their pain invisible, their cries inaudible, their membership in the human community implicitly denied. Speaking for them, from them, to them and through them, Du Bois, Agee, Olsen, and Anzaldúa brought to a broad general audience dimensions of experience previously absent from American newspapers, magazines, and books. The formal experiments they embraced were the sort of thing modernist poets tended to fool with: these four writers pressed them into the service of nonfiction. The truths they had to tell concerned subjects routinely dismissed as too boring or dull by the nation's journalists: these writers told those truths with poetry and passion usually reserved for fiction. Their agenda was clear: make the reader feel what you have felt, even if you have to break rules, customs, and conventions to do so. The passionate cultural reports they produced transcended existing paradigms and stretched the boundaries of our culture in enormously rich and fruitful ways. They also helped map new directions and set new standards for literary nonfiction in the twentieth century.

American literary history is punctuated by the appearance, from time to time, of literary experiments that transform the discourse of which they are a part. W. E. B. Du Bois's *The Souls of Black Folk,* James Agee's *Let Us Now Praise Famous Men,* Tillie Olsen's *Silences,* and Gloria Anzaldúa's *Borderlands/La Frontera* are four such transformative works.

Each book grew directly out of its author's struggle with a particularly problematical writing challenge—problematical for personal, political, and literary reasons. Du Bois, Agee, Olsen, and Anzaldúa found that generic conventions and available narrative forms could not do justice to their experiences and their thoughts. Indeed, each writer had grave doubts about translating his or her thoughts on the subject at hand into print at all. What initially appeared as the problem, however, ultimately proved to be the solution: by foregrounding rather than burying the records of *how* they grappled with their doubts and ambivalences, these writers produced texts that were both unique and groundbreaking—and which, in turn, would pave the way for many others. These writers' perceptions of the limitations of the generic conventions propelled them to break through those conventions in particularly innovative ways. Their experiments, in turn, made space for new voices— voices previously unheard, for the most part, in American nonfiction narrative.

The boldly original experiments they created spotlighted the difficulty of the challenges they faced and also emphasized the creativity that would enable them to meet them. Du Bois pasted bars of musical notation across the page. Agee inserted a line of pure punctuation into his text. Olsen experimented with typography, chapter length, structure, and layout. Anzaldúa constantly switched back and forth from one linguistic code to another within a single paragraph or sentence. Each of these disparate strategies was shaped by some common goals. All four writers wanted to disrupt patterns of perception familiar to the reader. They wanted to defamiliarize the familiar, explode conventional expectations, break down the reader's sense of equilibrium, surprise, challenge, and throw the reader off guard. In short, they wanted their readers to approach the text in ways that had never been required of them before, and to be changed, profoundly, in the process.

I would like to discuss some of the often daring, supremely self-conscious, sometimes bizarre, and never replicated formal experiments that help make the texts under discussion distinctive. In each case I will briefly allude to the generic conventions each work tries to subvert before exploring the particular strategy of subversion and the reasons behind it.

W. E. B. DU BOIS

The universe of discourse Du Bois's 1903 work, *The Souls of Black Folk,* would enter—nonfiction writing on race in America—was one in which Du Bois initially felt comfortable, since he began to shape it himself in the 1890s in key ways. Rather than struggling against the limitations of conventions defined by others, Du Bois would find himself—for personal, political, and literary reasons—pushing against conventions that were, to a large extent, of his own making. The results were radical and groundbreaking.

While nonfiction writing on race in America goes back as far as writing on America itself,[1] nonfiction accounts of black life in America written by blacks themselves increased dramatically in the nineteenth century as black newspapers, periodicals, and book-length narratives proliferated. Each of these forms played a role in Du Bois's apprenticeship, and each would shape *The Souls of Black Folk* in key ways. The first black newspaper in America, *Freedom's Journal,* appeared in 1827; by 1890 there were more than 150 black newspapers and magazines published in this country.[2] Du Bois entered the world of African-American journalism as a teenager, when he became the western Massachusetts correspondent for T. Thomas Fortune's weekly *New York Globe.*[3] In college, at Fisk University, he served on the editorial board of the *Fisk Herald,* a monthly, and was its editor-in-chief his third year.[4] Throughout the nineteenth century blacks were rarely on the editorial or reportorial staffs of mainstream white publications, where coverage of black life was most often limited to the police blotter[5]; however, a few mainstream national magazines ran occasional contributions from black writers. In the 1890s, Du Bois contributed memorable articles on African-American life to publications that reached predominantly white audiences, such as *The Atlantic Monthly* and the *Dial.*[6] Several of these pieces—such as "Strivings of the Negro People," "The Freedman's Bureau," "A Negro Schoolmaster in the New South," and "Of the Training of Black Men," all of which appeared in *Atlantic Monthly,* or "The Evolution of Negro Leadership," which ran in *Dial*—would find their way, virtually unchanged, into *The Souls of Black Folk.* In many ways *Souls* would be an extended cultural report—a report of a journey south, into the Black Belt, by a northern black to whom that journey was both different and new.[7] But despite its antecedents in Du Bois's journalism *Souls* was more than a series of articles stitched together: the book-length narrative Du Bois published in 1903 was infinitely larger than the sum of its parts.

Before the 1890s book-length nonfiction narratives of black life in America by black Americans could generally be divided into two categories: slave narratives—such as those of Frederick Douglass, Henry Bibb, William Wells Brown, Solomon Northrup, and Harriet Jacobs—and histories—principally those by William Stills, Joseph T. Wilson, and George Washington Williams.[8] The 1890s saw the rise of more theoretical and analytical works such as Anna Julia Cooper's *A Voice from the South* and Ida B. Wells-Barnett's statistical survey of lynchings, *A Red Record.*[9] Until the publication of Du Bois's *The Philadelphia Negro* in 1899, however, the methods of social science had not been systematically applied to the black community. With this book, and with the series of Atlanta University studies that followed it under his supervision, Du Bois created a tradition of sociological investigation of black life in America.

Du Bois left his teaching position at Wilberforce to undertake research, under the auspices of the University of Pennsylvania, on the black community in Philadelphia. It was absolutely clear to him at the time that such an investigation was the best way to address what he called "The Negro problem." Later, looking back on that moment in time, he wrote "The Negro problem was in my mind a matter of systematic investigation and intelligent understanding. The world thinking was wrong about race, because it did not know. The ultimate evil was stupidity. The cure for it was knowledge based on scientific investigation."[10] Du Bois himself, armed with the tools of investigation he had acquired from outstanding teachers at Fisk, Harvard, and Berlin, "determined," as he put it, "to put science into sociology through a study of the condition and problems of my own group."[11] The result was *The Philadelphia Negro,* a remarkable and historic study that broke new ground by offering "environmental rather than racial explanations of his subject."[12] When his book came out, he included in it a plea that funding be made available for further studies of this sort. "Without a doubt," he felt, "the first effective step toward the solving of the Negro question will be the endowment of a Negro college which is not merely a teaching body, but a center of sociological research."[13]

When Du Bois was put in charge of what he envisioned as a vast series of such studies at Atlanta University, he had high hopes and high standards for the research they would yield; despite the almost proselytical zeal with which he championed the cause of sociological research, he was determined to be a sober scientist every inch of the way: "I was going to study the facts, any and all facts, concerning the American Negro and his plight," Du Bois later recalled, "and by measurement and comparison and research, work up to any valid generalization which I could. I entered this primarily with the utilitarian object of reform and uplift; but nevertheless, I wanted to do the work with scientific accuracy."[14] The published studies that resulted explored such issues as black mortality, urbanization, efforts of blacks for their own social betterment, and so on. They were highly regarded nationally and internationally and Du Bois found himself in great demand as a speaker.

At that point, however, an incident occurred that made Du Bois doubt the worth of such projects: something happened that made him feel that the dry sociology he had been writing and overseeing was not just inadequate, but also irrelevant. Du Bois described the crisis himself in this way:

At the very time when my studies were most successful there cut across this plan which I had as a scientist, a red ray which could not be ignored. I remember when it first, as it were, startled me to my feet. A poor Negro in central Georgia, Sam Hose, had killed his landlord's wife. I wrote out a careful and reasoned statement concerning the evident facts and started down to the Atlanta Constitution office, carrying in my pocket a letter of introduction to Joel Chandler Harris. I did not get there. On the way news met me: Sam Hose had been lynched, and they said that his knuckles were on exhibition at a grocery store farther down on Mitchell Street, along which I was walking. I turned back to the University. I began to turn aside from my work.[15]

Du Bois concluded that "one could not be a calm, cool, and detached scientist while Negroes were lynched, murdered and starved."[16] Du Bois questioned the use of sociological studies in such an atmosphere of violence, ignorance, and betrayal. It dawned on him that white America failed to see that the same human qualities that animated *it* also animated those whose skin was another color: white America did not even recognize black people as human. It became clear to Du Bois that collecting more statistics was not the way to correct such chilling myopia; nor, for that matter, was writing more articles for the *Atlantic*. Something different was required. That "something different" for Du Bois would turn out to be his seminal 1903 work, *The Souls of Black Folk*.[17] A mixture of journalism, social science, autobiography, fiction, poetry, musicology, and history, *The Souls of Black Folk* resembled nothing that had come before. It sought new ends through new means, and it would change forever the nature of American writing on race.

White America failed to grant black America basic human rights because it denied black America's basic humanity. How could a writer execute the monumental task of conveying to his readers the depth and depravity of their own ignorance and blindness? Du Bois employed many strategies toward this end; the one that will concern us here is the presence of the epigraphs and bars of music with which Du Bois begins each chapter (See Figure 6.1).

If the products of "culture"—such as literature, sculpture, and architecture—were evidence of membership in the human community upon which everyone could agree, what could one find to present as "culture" from a people who had been forbidden to read or write, to sculpt or build? The "sorrow songs," as Du Bois called them, or the spirituals sprang to mind as one cultural artifact through which vast continents of pain and emotion were refracted. But how does one turn a "sorrow song" into "culture" by conventional standards? Du Bois solved the problem by foregrounding—rather than burying—the difficulty of blending two yardsticks of spiritual elevation—"sorrow songs" and famous works of the Western literary tradition.

Du Bois lays out the plan of his book for his reader in the "Forethought" that introduces the volume.

> Herein lie buried many things which if read with patience may show the strange meaning of being black here in the dawning of the Twentieth Century. This meaning is not without interest to you, Gentle reader; for the problem of the Twentieth Century is the problem of the color line.
>
> I pray you, then, receive my little book in all charity, studying my words with me, forgiving mistake and foible for the sake of the faith and passion that is in me, and seeking the grain of truth hidden there . . . I have sought here to sketch, in vague uncertain outline, the spiritual world in which ten thousand Americans live and strive.
>
> . . . Leaving, then, the world of the white man, I have stepped within the Veil, raising it that you may view faintly its deeper recesses,—the meaning of its religion, the passion of its human sorrows, and the struggle of its greater souls. . . . Before each chapter, as now printed stands a bar of the Sorrow Songs,—some echo of haunting melody from the only American music which welled up from black souls in the dark past. And, finally, need I add that I who speak here am bone of the bone and flesh of the flesh of them that live within the Veil?[18]

Several points about the "Forethought" are worth emphasizing here. First of all, the reader is warned that the "meaning" of being black at the dawn of the twentieth century is "strange" yet relevant to the reader. The reader is warned that the material about to be encountered may transcend familiar categories of understanding, but must be assimilated nonetheless, for it bears on "the color line," the problem that will be "the problem of the Twentieth Century." What is the "color line," anyway? The fact that Du Bois does not feel obligated to define the term here suggests that his readers must feel they understand it. It clearly refers to the segregation (and implicit hierarchical separation) of races, a doctrine largely unchallenged in the dominant culture.

Du Bois notes that he will attempt to delineate "in vague uncertain outline, the spiritual world in which ten thousand Americans live and strive"; he will also explore the "deeper recesses" of life on one side of the "color line"—the "meaning of its religion, the passion of its human sorrow, and the struggle of its greater souls." He will portray this spiritual world, in part, with "a bar of the Sorrow Songs." What he does not say in his forethought is that every time "a bar of the Sorrow Songs" appears, it will be paired with a quotation from one of the familiar works of white high culture—that the two "quotations" will coexist on equal footing at the start of each chapter—and that while appearing to be from bizarrely different realms, they will prove to be addressing the same themes in the same ways. What he does not warn the reader is that these paired "quotations" will, by their very juxtaposition, reveal the notion of a "color line" to be a fiction—when it comes to the strivings of the human spirit or the aspirations of the human soul. It is a bold and daring experiment—all the more so in a book that humbly purports to lay out its plan straightforwardly from the start. And it works.

Despite Du Bois's warning that he will include bars of music in his book and despite his identification of what they are, the reader who turns the page to his first chapter confronts a text that is jarring, and, most definitely strange. Two verses from British poet Arthur Symons beginning, "O water, voice of my heart, crying in the sand," followed by three measures of unidentified notes and staff lines. The pairings that begin the other chapters are equally jarring and strange; there bars of unidentified music are paired with quotations from "Lowell," "Byron," "Whittier," "Omar Khayyam," "William Vaughn Moody," "Mrs. Browning," "Fiona Macleod," "Swinburne," and "Tennyson." One pairing is with a quote from the Bible ("Song of Solomon"). Another—perhaps the most striking juxtaposition of all—is with an untranslated German poem by Schiller.[19] All of the poems deal with human pain, suffering, aspirations, spiritual strivings—themes Du Bois addresses at length in his text.

Between the "Forethought" and his final chapter, "The Sorrow Songs," Du Bois makes no comment on the nature of the bars of music nor on his reason for including them. This silence, on Du Bois's part, requires readers to make sense of the material themselves—it demands that the reader forge a connection between the anonymous bar of music and the signed poem; it asserts, implicitly, that the bar of music is as worthy of the reader's attention as is the poem, even if the music's "meaning" seems, temporarily, at least, undecipherable. Is it so difficult, Du Bois seems to ask, to imagine that the phrase *"Die nicht fühlen, die nicht weinen!"* ("the tough-hearted and tearless")[20] inhabits the same moral universe as a black spiritual that mourns the toughened heart and unshed tears of generations of slaves? By leaving the precise "meaning" of the notes on the page indeterminate, Du Bois requires us to create meanings on our own. If we have difficulty "reading" the "sorrow songs" (a difficulty that an illiterate slave who did not know a word of German would never have), we are reminded that there are dimensions of human experience and culture that may not be decipherable in conventional ways, and that the universal yearning of the human soul toward freedom may express itself in a range of forms—only some of which may be "read" by those of us insulated by education, privilege, and opportunity.

Du Bois himself, as he well understood, must have seemed an impossibility, a monstrosity, in some ways, to many of his countrymen. He was a pupil of James and Santayana, the product of a German university and two American ones, who was gifted in languages, history, and science; his very existence was an affront to the hierarchical rankings of intelligence and competence presupposed by the "color line." As he yoked fragments of black musical tradition and Western literary tradition together in his book, so he yoked the two traditions together in his person: fluent in the language of Schiller, he was also fluent in the language expressed by the distinctive bars of "Nobody Knows the Troubles I've Seen." His conviction that his fellow Americans needed to be as conversant in both modes of discourse prompted the eloquent exercise in translation that is the final chapter of *Souls,* "The Sorrow Songs."

Here Du Bois finally "translates" for his readers the significance of much that has gone before.

Little of beauty has America given the world save the rude grandeur God stamped on her bosom; the human spirit in this new world has expressed itself in vigor and ingenuity rather than in beauty. And so by fateful chance the Negro folk-song—the rhythmic cry of the slave—stands today not simply as the sole American music, but as the most beautiful expression of human experience born this side the seas. It has been neglected, it has been, and is, half despised, and above all it has been persistently mistaken and misunderstood; but not withstanding, it remains as the singular spiritual heritage of the nation and the greatest gift of the Negro people.[21]

Du Bois then proceeds to tell the story of the Fisk Jubilee Singers, enterprising singers of sorrow songs who, some thirty years before returned from a seven-year singing tour with the then-staggering sum of a hundred and fifty thousand dollars—with which they founded Fisk. (It is ironic, of course, that the university that first exposed Du Bois to the great works of Western culture came into being—quite literally—because educated whites were willing to pay handsomely to be exposed to the black folk tradition Du Bois champions here.)

Du Bois then asks the question his readers must have asked at the start of each chapter: "What are these songs, and what do they mean?"[22] He devotes the next nine pages—the final pages of the book, to his answer. Avowing from the start that he knows "little of music and can say nothing in technical phrase," Du Bois proceeds to explore the symbolic significance of the songs—the sorrow songs as complex cultural artifact. "I know that these songs are the articulate message of the slave to the world."[23] As a self-conscious corrective to the myth of the presence of song as evidence of the "careless and happy" quality of antebellum life, Du Bois asserts that the songs "are the music of an unhappy people, of the children of disappointment; they tell of death and suffering and unvoiced longings toward a truer world, of misty wanderings and hidden ways. The songs are the siftings of centuries; the music is far more ancient than the words, and in it we can trace here and there signs of development." He quotes in full an African melody his great, great grandmother sang and passed on to her children; the melody is two hundred years old.[24] He then traces echoes of that African "forest of melody" in the ten master songs from which the fragments at the head of each chapter have been taken. The words, he notes, "conceal much of real poetry and meaning beneath conventional theology and unmeaning rhapsody." He goes on to explicate some of the images and their significance. While the bars of music appeared as floating fragments of an unknown tongue at the chapter heads, they are here attached to words, to songs, to families of melodies, to history, to culture, and to the experience of a people. In short, just as the educated reader would have been likely to know that a fragment from "Mrs. Browning" had a recognizable place in a tradition many hundreds of years old, the reader has been educated by this point to understand that the fragments from the sorrow songs occupy an analogous place in an analogous tradition. Du Bois writes:

> Through all the sorrow of the Sorrow Songs there breathes a hope—a faith in the ultimate justice of things. The minor cadences of despair change often to triumph and calm confidence.
> Sometimes it is faith in life, sometimes a faith in death, sometimes assurance of boundless justice in some fair world beyond. But whichever it is, the meaning is always clear: that sometime, somewhere, men will judge men by their souls and not by their skins. Is such a hope justified? Do the Sorrow Songs sing true?[25]

In one sense, all that has gone before, up to this point, is prologue. For it is in the concluding page and a half of *Souls* that Du Bois states his central thesis and makes his most impassioned plea. The eloquence and directness of his concluding statements make the oblique route he has taken to lead up to this point seem all the more strategic and brilliant:

> The silently growing assumption of this age is that the . . . backward races of to-day are of proven inefficiency and not worth the saving. Such an assumption is the arrogance of peoples irreverent toward Time and ignorant of the deeds of men. A thousand years ago such an assumption, easily possible, would have made it difficult for the Teuton to prove his right to life. Two thousand years ago such dogmatism, readily welcome, would have scouted the idea of blond races ever leading civilization. So wofully unorganized is sociological knowledge that the meaning of progress, the meaning of "swift" and "slow" in human doing, and the limits of human perfectability, are veiled, unanswered sphinxes on the shores of science . . .
> Your country? How came it yours? Before the Pilgrims landed we were here. Here we have brought our three gifts and mingled

them with yours: a gift of story and song—soft, stirring melody in an ill-harmonized and unmelodious land; the gift of sweat and brawn to beat back the wilderness, conquer the soil, and lay the foundations of this vast economic empire two hundred years earlier than your weak hands could have done it; the third, a gift of the Spirit. . . . Actively we have woven ourselves with the very warp and woof of this nation,—we fought their battles, shared their sorrow, mingled our blood with theirs, and generation after generation have pleaded with a headstrong, careless people to despise not Justice, Mercy and Truth lest the nation be smitten with a curse. Our song, our toil, our cheer, and warning have been given to this nation in blood-brotherhood.[26]

Just as his reader is likely to find Du Bois's practice of yoking together the bars of music and verses of the poets at the start of each chapter foreign and strange, so too, Du Bois seems to say, America has found the notion of black and white yoked together in "blood-brotherhood" to a common fate and common destiny foreign and strange. Yet the two worlds—separated only by gulfs of understanding, not by what lies within—are really one; they are linked indivisibly, one to the other. Only a change in consciousness, Du Bois seems to feel, can bring about a real change in material conditions: the white world must learn to recognize the cultural expressions of the world behind the veil as those of its "blood-brothers," sharing common spiritual strivings and a common soul; by the same token, as Du Bois emphasized in his writings on "the talented tenth" and elsewhere, the black world must come to see the cultural expressions of the worlds beyond the veil—traditions of Western lyric poetry, for example—as potential instruments of change, as forms black writers should feel free to appropriate and make their own. Du Bois was not asking simply that whites tolerate the cultural expression of blacks; he made it clear that their making an effort to *understand and appreciate* that expression was essential to America's survival as a whole and healthy nation. In this he was not asking simply that blacks imitate white cultural forms; he was asking that they combine their awareness of the usefulness of those forms with the special strengths of their own personal heritage to make something new. As he had said in a speech in 1897, the destiny of black people in America was not absorption into or "a servile imitation of Anglo-Saxon culture, but a stalwart originality."[27]

In *The Souls of Black Folk* Du Bois validates the folk culture in which this pain and striving have been expressed—and gives readers the skills as well as the motive to "read" it on their own. His example (on this particular issue, as well as many others) has served to inspire a tremendous part of contemporary African-American literary and critical traditions.

JAMES AGEE

In 1936, when James Agee set out to do a story with Walker Evans on Alabama tenant farmers, at the direction of *Fortune* magazine, his editors—and perhaps Agee himself—expected him to be able to turn out an article of the sort that would fit into the "Life and Circumstances" series *Fortune* had been running on poor and lower-middle-class Americans. What they got, of course, was nothing of the sort. Agee's article, they asserted, was unprintable.

The problem was most certainly not that *Fortune*'s expectations were vague or ill-defined. Agee's article was to follow the pattern of the documentary series that the magazine had been running, and be "a photographic and verbal record of the daily living of an 'average,' or 'representative,' family of white tenant farmers."[28] His task was to describe the tenant farmers' "daily life and its seasonal cycles, to report on the farm economy, on the efforts of the federal and state governments to improve the situation, on the theories held by Southern liberals and on the history of the two farm associations that had been established."[29] The tone of such pieces was one of "breezy condescension"[30]; the poor people profiled were described in a manner designed to amuse or entertain *Fortune*'s affluent, powerful readers.

The universe of discourse that characterized series such as this one was, of course, that of documentary nonfiction, an increasingly popular and characteristic form of expression in the 1930s. The varieties of the 1930s documentary—which included nonfiction film, popularized case-worker studies, social science writing, documentary reportage, and documentary books—have been eloquently analyzed by William Stott in his magisterial work, *Documentary Expression and Thirties America* (1973).

Perhaps the most useful way of summarizing some aspects of the genre would be to describe one documentary book that typified the form for so many Americans, a book that happens to deal with the same subject as Agee's and Evans's book, Southern tenant farmers, and that, like theirs, was begun in the summer of 1936.

Margaret Bourke-White and Erskine Caldwell's highly acclaimed 1937 work *You Have Seen Their Faces,* in which the plight of the Southern tenant farmer was explored through text, photo, and caption, has come to typify for many the 1930's documentary book. Critics at the time hailed it as "the first volume to be presented in a new genre,"[31] but as William Stott has shown, it had many precedents. Foremost of these was "the 'photo essay' in the Luce magazines, for which Bourke-White had worked since 1929, and, specifically, such features as *Fortune*'s 'Life and Circumstances' series. . . . These articles balanced text and photos as the documentary books later did, and to the same end: to make vivid the lives of people on society's lower rungs."[32] As Stott notes, "the captions, like the rest of the book, reduce the lives and consciousness of the tenant farmers to force the audience to pity them. The subjects say either, 'Look at me, how wretched my life is,' or 'Look at me, my life is so wretched I don't even know it.' "[33] The book— and particularly its most sentimental features—garnered tremendous critical acclaim. One leading critic effused, "the quotations printed beneath the photographs are exactly right; the photographs themselves are almost beyond praise."[34]

The book was a huge commercial as well as critical success. As Walker Evans later commented, he and Agee—who were still two years away from completing their book when the Caldwell–Bourke-White book came out—were furious. They felt that Caldwell and Bourke-White had both profited from and exploited their subjects—people who, as Evans put it, "had already been so exploited." What particularly galled them was that they felt Caldwell and Bourke-White were exploiting "them without even knowing that that was what (they) were doing."[35] Indeed, I believe, it was Agee's and Evans's sensitivity to the whole issue of exploitation that, in the end, enabled them to produce a volume that transcended the limitations of the documentary genre. In the case of the text, in particular, Agee's decision to foreground and directly confront the anxiety and guilt he felt about "using" the painfully constricted lives of the tenant farmers as the subject of his book would prompt some of his most daring, dramatic, and successful experiments in narrative. (It was not, however, a success when it was published. While the Caldwell–Bourke-White book became a best-seller, the Agee–Evans book was, as one critic put it, "one of the most spectacular publishing failures" of the year when it appeared.[36] Critics were typically brutal; sales were miniscule.[37] Only after its republication in 1960 would the book come into its own.)

Agee must have suspected fairly early in that summer of 1936 that he would be unable to write the kind of piece that *Fortune* expected of him. When the trip was over he would write to Father Flye, "the trip was very hard, and certainly one of the best things that ever happened to me. Writing what we found is a different matter, impossible in any form *Fortune* can use."[38] Agee was himself profoundly moved—and changed—by the experience of becoming a part of the tenant farmers' lives. He lacked the detachment, condescension, and self-confidence necessary to produce a typical "Life and Circumstances" article. If he was certain that he could not write the article he had been assigned to write, however, he quickly came to doubt both the possibility, and indeed the morality, of his writing about these people at all. In what ways, he asked himself, did his very presence as a writer constitute an act of immense arrogance and presumptuousness? He fantasized about ways of evoking and paying homage to their lives without words—with "bits of cotton, lumps of earth," and so forth. It was only a fantasy: his medium, as he grimly came to accept, had to be words. He grew more and more nervous, more uncomfortable—not with the people, whom he found himself, quite literally, coming to love, but with his assignment. From this intense personal crisis an intensely personal book would emerge.

Out of Agee's excruciatingly honest and open struggle with his own responses to the world of the sharecroppers came something new. His painstaking attention to the quotidian details of lives generally thought by the reading public to be too insignificant for such loving scrutiny—combined with the painfully honest scrutiny to which he subjected himself in the act of making these observations—produced one of the most startling texts contemporary American culture has produced.

When Agee returned to his notes for the book on the tenant farmers in 1938, the project had clearly changed from what it was at its start; however, what it had changed into was far from clear. As Agee's biographer Genevieve Moreau notes:

> He now envisioned it both as an autobiographical narrative present-ing its author as an ignorant and unworthy journalist confronting an absurd and unjust system and as a meditative poem based on the sociological materials of the situation of the working poor in Ala-bama. On a level deeper than his anger at the oppression of a social class, his denunciation of abuses and defense of the oppressed, Agee wanted to demonstrate that the very issue of human existence was involved. It was necessary, in the end, to place the Southern tenant farmer in the wider context of humanity as a whole.[39]

To accomplish this goal,

> an infinity of possible forms suggested themselves to him: social tract, journal, realistic narrative, documentary, poem of cosmic scope. He ultimately decided he could not limit the work to any one particular form; instead, he attempted to integrate all the possibili-ties into a single narrative that would make use of each within a unified whole. He was determined to achieve a new art form by ignoring the outmoded rules that these forms of literature followed. At the same time he also intended to defy the conventions of *art* and *literature* in the commonly accepted sense of those words.[40]

Agee's challenge, as he would put it in *Let Us Now Praise Fa-mous Men,* was "to recognize the stature of a portion of unimag-ined existence, and to contrive techniques proper to its recording, communication, analysis and defense."[41] (The challenge was not unlike that which Du Bois took on in *Souls,* in which his validation of "the stature of a portion of unimagined existence" similarly prompted him to experiment with new "techniques proper to its recording, communication, analysis and defense.") But the tradi-tional repertoire of options open to the writer seemed strikingly thin, inadequate, limiting, and artificial—even at best destined to yield only vague, tentative approximations of the experience the writer wanted to evoke.

Agee wished he could dispense with words and paper altogether: "If I could do it," he tells us, "I'd do no writing at all here. It would be photographs; the rest would be fragments of cloth, bits of cotton, lumps of earth, records of speech, pieces of wood and iron, phials of odors, plates of food and of excrement."[42] Dazzled and dazed by "the cruel radiance of what is," Agee stressed that *everything* about the sharecroppers' world mattered to him *immensely:*

> Ultimately, it is intended that this record and analysis be exhaustive, with no detail, however trivial it may seem, left untouched, no relevancy avoided, which lies within the power of remembrance to maintain, of the intelligence to perceive, and of the spirit to persist in.
>
> Of this ultimate intention the present volume is merely portent and fragment, experiment, dissonant prologue.[43]

Here we have a clue to the insight that enabled Agee to write at all: his awareness that his book would be "true" only to the extent that it acknowledged its own incompleteness, that a measure of his success would be the extent to which he made his reader aware of his *failure* to cast the sharecroppers' experience into words. Agee becomes even more specific on this point:

> This is a *book* only by necessity. More seriously, it is an effort in human actuality, in which the reader is no less centrally involved than the authors and those of whom they tell. Those who wish actively to participate in the subject, in whatever degree of understanding, friendship, or hostility, are invited to address the authors in care of the publishers.[44]

It is ironic, perhaps, that what is offered as an assertion of weakness, limitation, and humility would eventually become—for the reader—an impressive sign of strength, scope, and confidence on the part of the author. For, as Agee himself asserts these qualities before our eyes, we come to understand the strength it takes to admit weakness, the scope of vision required to acknowledge one's limits, and the supreme confidence that one must have to invite the reader, as an equal, to continue an ongoing dialogue with the author.

How does Agee pull all this off? Here I will examine one intriguing line of the book which is, for me, emblematic of Agee's larger strategy and characteristic of what makes this book so different from all that went before it. It is a line that stops his readers dead in their tracks and forces them to look at the text at hand—and the world it so humbly approximates—in new ways. While on one level it is an expression of the writer's basic humility toward his endeavor, on another level it is also a daring and defiant assertion of his mastery of his craft.

Broken up as the book is into such a diverse collection of different narrative enterprises, one would almost begin to expect the unexpected and be immune to surprise. Yet one *is* surprised when on p. 76 of the text one encounters a line of pure punctuation:

$$(\qquad (?) \qquad) \qquad :)$$

It is ironic that James Agee, who felt himself so completely estranged from abstract formalism and so completely embedded in the concrete world of the photographer, was responsible for one of the most abstract formulations of human destiny ever written.

The line, which comes just before, and indeed introduces, the famous "How were we caught?" passage, etches in a set of abstract symbols the boundaries that circumscribe the sharecropper's life (see Figure 6.2). The all-inclusive larger parenthesis suggests the life as a whole, limited in both years and in the direction and path the years may take. The questioning that intrudes ("How were we caught?" "In what way were we trapped?") is itself parenthetical—it does not affect the course the life takes in any significant way. Neither does the empty anticipation of death which comes after the active years of life are over. Those who have "no memorial; who perished, as though they had never been; . . . and their children after them," are, in the text of human history, a subordinate and deletable clause.[45] What better way to evoke this destiny than through the abstract symbol for a subordinate and deletable clause, the parenthesis?

The "How were we caught?" section of the book—in which Agee fabricates thoughts and presents them as if the tenant farmers had expressed them—made Walker Evans squirm, and has incurred the distaste of critics since the book first appeared. Even critics as deeply supportive of Agee's enterprise as William Stott have found this section to be offensively "arrogant."[46] Such a deception *is* arrogant. Indeed, I would claim, Agee knew it. Despite his professions of humility before the men and women he encountered, Agee recognized that the form into which he would cast these lives would be, ultimately, his and not theirs. Perhaps this line of punctuation, which so directly parallels the train of thought in the "How were we caught?" sequence is Agee's implicit acknowledgment of the invented, projected, and ultimately conjectural nature of his book—and, perhaps, of any possible structure of words into which he might attempt to encompass the sharecroppers' lives.

Agee recognized that despite his professions of respect for the concrete solidity of the sharecropper's world, the world the reader would receive would be not that of the tenant farmers but the world of words woven by the author.[47] Was his world of words, like the world of the sharecroppers, destined to have that curious status all books are destined to have in the course of things—that of something deletable from the course of history, which exists in that parenthetical "other" world of letters, but leaves the actualities of the "real" world out beyond it unchanged? Just as Agee refutes throughout the book (from its ironic title, to its end) the notion of the sharecropper's life as parenthesis, he also refutes the notion of his own enterprise as parenthesis. For all his protestations to the contrary, his text manages to be, among many other things, a celebration of his own power as a writer. So what if the man is tentative, insecure, self-effacing, confused, diffident, and embarassed? The artist in him transmutes that insecurity and diffidence into prose that transcends itself, that startles with its candor and creativity, and that jolts readers out of familiar patterns of perception and rhythms of reading and forces them to confront bare, forceful, autonomous imagination in the form of a line of pure abstract symbols. On the one hand the line of pure punctuation reminds the reader of the ultimate aridity, arbitrariness, and artificial nature of the printed page. On the other hand, defining, as it does, its opposite, it also reminds the reader of the extraordinary fullness and richness the artist's imagination has been able to impart to an enterprise that is, after all, only smears of ink on expanses of paper.

TILLIE OLSEN

At first glance, Tillie Olsen's book *Silences,* published in 1978,[48] would seem to inhabit that readily identifiable universe of discourse known as literary criticism. But ponder the significance of the term *literary criticism* for a moment. It refers to the practice of writing thoughtfully about an existing body of texts: it presumes a body of "literature" about which one can wax "critical." In this remarkable volume, however, Tillie Olsen is only partially concerned with writing critically about a body of existing literature. Her larger concern is with the books that never got written and with the people who never wrote them. Like Du Bois and Agee before her—both of whom were very much on her mind as she wrote *Silences*[49]—Olsen cares passionately about documenting the experiences of those who cannot tell their own stories. Bearing witness to realities that have heretofore eluded the printed page, she wants to issue a cultural report from the realm of the silent and the silenced. Olsen also wants to understand patterns of silencing that persist into the present. She wants to write not only about the books that have never been written, but about the difficulty—for many would-be writers—of writing books even now.

How do you write literary criticism about the lack of a body of literature to criticize? And how do you effectively address the nearly insurmountable obstacles facing, say, working-class women writers, without creating a book that is, by its very existence, close to being self-refuting?

For Tillie Olsen herself these questions were preceded by nearly twenty years of personal silence as a writer. Her own experience during the years between the publication of her early journalism and fiction in the 1930s and the appearance in 1961 of her short-story collection, *Tell Me a Riddle,* fueled her interest in the contexts of other silences she perceived on the part of other writers. She was not, of course, the first writer to address the notion of silences, particularly on the part of women. Virginia Woolf had broached the subject in 1928 in *A Room of One's Own,* which still stands as the landmark work exploring the preconditions for productivity and creativity on the part of the woman writer. But if Woolf had laid the issues on the table in 1928 in England, no American had chosen to embrace them as her principal concern before Olsen did in 1965.[50] If her magazine article titled "Silences" did nothing more than try to broach this subject in terms suited to its exploration in America, it would still be truly groundbreaking; but Olsen went even further.

The notion—now so generally accepted as to seem almost a

truism—that a woman's personal experiences could be directly relevant to her insights as a critic was as foreign to the world of criticism in 1965 as was the concept of analyzing "silences." The "New Criticism," of course, was by then solidly enthroned. The text itself—and the text most definitely *by* itself—was confidently championed as the proper object of criticism. Against this backdrop, Olsen's concerns not only with the lack of texts, but also with the details of the lives of those who failed to produce texts, stands out as particularly bold and subversive.

In 1965 Olsen had the field virtually all to herself; by 1978, when her book *Silences* appeared, a host of writers had acknowledged in print the debt they owed her.[51] Despite the fact that feminist criticism would become, in the late 1960s and 1970s, a field heady with experimentation, I would argue that even against that backdrop, Olsen's book *Silences* still stands out. No other feminist text from this period (with the possible exception of Mary Ellmann's remarkable 1968 work *Thinking About Women*) is characterized by such dramatically new formal and technical innovations—particularly by innovations as uniquely suited to the subject that is being explored.

What is the relationship of a writer with Olsen's agenda to the traditions of discourse into which her book of necessity enters? In a word, complex. In this essay there is space to focus on only one of the techniques Olsen employs to achieve her end. I would argue, however, that the entire catalog of innovative and creative strategies she uses in the book are designed to teach her readers to allow voices previously faint or silent (including, one might add, the reader's own) to be heard. She is remarkably successful in this endeavor.

How does one learn to "hear" silence? Explicating this apparent oxymoron is indeed the principal technical challenge Olsen faces. It might help to remember that the word *silence,* for Olsen, is itself, for the most part, a metaphor for the absence of written texts. The direct analogue to *silence* in a visual context is *blankness.* Olsen sagely recognizes that the auditory concept of *silence* is more powerful and allusive than the visual concept of *blankness* for a variety of reasons, not least of which is the convenient link with the familiar silencing of women's real voices in family conversations, as in the larger cultural conversations that have gone on throughout history. But hearing silence—indeed investigating silence—is by definition problematical. Here I would like to focus on one technique Olsen employs to help her readers learn to do so—a technique whereby she subtly retreats from her auditory metaphor and returns to the realm that really concerns her, the visual: the books that are not there, the written words that are absent; in short, the blankness that gapes at Olsen from the pages of literary history (see Figure 6.3).

Olsen's decision to follow the more conventional, straightforward critical narrative of the first 118 pages of the book ("Part One—Silences") with the collection of fragments that characterizes the second half of the book ("Part Two—Acerbs, Asides, Amulets, Exhumations, Sources, Deepenings, Roundings, Expansions") is both strategic and effective. Olsen dramatizes the theme of "silence" with peculiar immediacy and intensity by confronting the reader with the "visual silence" of blankness on the page. This experimental exploration of the theme of silence follows the more conventional examination of the theme in "Part One": the direct provides the context for the oblique. Thus it is to "Part One" that we must first direct our attention—or rather, to the pages that immediately precede "Part One."

After the "Acknowledgments"[52] comes the page usually reserved for the dedication, and Olsen's is there, where we expect it to be. But Olsen has three pages reserved for dedication (instead of the usual one), pages in which the ratio of blank space to print steadily decreases.[53]

The first page contains a two-line quote from Thomas Hardy; the rest of the page is blank.

> (This) is sent out to those whose souls the iron has
> entered, and has entered deeply at some time of their lives.
> —Thomas Hardy, of his
> *Jude the Obscure*

The next page contains two paragraphs surrounded by an otherwise blank page:

> For our silenced people, century after century their beings consumed in the hard, everyday essential work of maintaining human life. Their art, which still they made—as their other contributions—anonymous; refused respect, recognition; lost.
>
> For those of us (few yet in number, for the way is punishing), their kin and descendants, who begin to emerge into more flowered and

rewarded use of our selves in ways denied to them;—and by our achievement bearing witness to what was (and still is) being lost, silenced.

The third and final page of dedicatory/prefatory material spells out precisely what the author plans to do. It is necessary to quote from it at some length:

> *Literary history and the present are dark with silences: some the silences for years by our acknowledged great; some silences hidden; some the ceasing to publish after one work appears; some the never coming to book form at all.*
>
> *These are not natural silences, that necessary time for renewal, lying fallow, gestation in the natural cycle of creation. The silences I speak of here are unnatural; the unnatural thwarting of what struggles to come into being but cannot. In the old, the obvious parallels: when the seed strikes stone; the soil will not sustain; the spring is false; the time is drought or blight or infestation; the frost comes premature.*
>
> This book is about such silences. It is concerned with the relationship of cirumstances—including class, color, sex; the times, climate into which one is born—to the creation of literature. . . . This book is not an orthodoxly written work of academic scholarship.
>
> Do not approach it as such. Nor did it come into being through choosing a subject, then researching for it. The substance herein was long in accumulation, garnered over fifty years, near a lifetime; the thought came slow, hard-won; the talks and essay, the book itself, elicited.
>
> A passion and a purpose inform its pages: love for my incomparable medium, literature; hatred for all that, societally rooted, unnecessarily lessens and denies it; slows, impairs, silences writers.
>
> It is written to re-dedicate and encourage.
>
> *I intend to bring you joy, courage, perspicacity, defiance.*
> —André Gide

The words on Olsen's dedicatory/prefatory pages, first just two lines on an expanse of empty whiteness, gradually increase, like a crescendo, until the blank space is dominated by text. The crescendo of these pages announces the text that follows: the table of contents—impressive, extensive, organized, analytical topics that march confidently across the page, taking up whatever space there is—for this table is, after all, *this* author's triumph over those obstacles she will enumerate. The complex and subversive prefatory pages of acknowledgment and dedication have earned Olsen the right to parade her "Table of Contents" without apology or awkwardness; she has established the context in which this convention has new meaning. It is not just "a" table of contents—it is *her* table of contents, testimony to the strength, "joy, courage, perspicacity, defiance" with which it was produced.

It is in these succinct prefatory pages that Olsen establishes both the nature of her subject matter and the nature of the challenge she confronted when she dealt with it. She makes clear from the start her conviction that the fact that something has been undervalued, or lost, or refused respect in no way takes away its potential claim to being "art." (Echoes of Du Bois spring to mind.) She makes it clear that something may have value despite the fact that the arbiters of value count it as nothing; she lets us know that those factors that assign value to some human creations and not to others may be more arbitrary than we might think. The reader is directed to focus intently on "our silenced people" whose beings were "consumed in the hard, everyday essential work of maintaining human life." Much of what appears as silence in history, might, then, be heard if we listen in other than customary ways.

How does Olsen herself, this consummate writer who is capable of turning out not only the extensive and impressive "Table of Contents" but also the eloquent pages that follow, "connect" with the silent ones to whom, and about whom, she writes? By the same token, how does she get readers—more likely than not to be privileged, educated, already on their way to overcoming their own silence—to connect with those less fortunate than themselves? The answer is in her last phrase, in which she exhorts those few who are beginning to "flower" to bear witness to the lives of those who failed to flower: "by our very achievement bearing witness to what was (and still is) being lost, silenced."[54] It is a masterful embrace with which Olsen draws all of us into her circle of listeners. She writes for all those silenced, and for all those not silenced—and the achievement of those who are not silenced, such as that of Tillie Olsen herself—bears witness by its very existence to what might have been, in the shadows of what never was.

How do you address the enigma of writing a book about the books that have not been written? How do you cope with being so extraordinarily eloquent about the silencing of the potential eloquence of others? For Olsen the answer is "without apology"—but with empathy, sensitivity, warmth, and constant awareness of the tightrope one is walking. "Do I stun or impress?" she might ask, in a tone Walt Whitman might have used. "Very well, then, I stun or impress—and so might the generations silenced by work, care, constricted vision. What I can do they might have done. I am nothing special," Olsen seems to say, "I merely bear witness to what was lost when they were silenced or ignored."

"Part One—Silences," which takes up slightly less than half of the volume, consists of three fairly conventionally structured essays that explore Olsen's central theme from three different angles. The headnote that introduces the first essay, "Silences in Literature" (the original version of which ran in *Harper's* in 1965) makes clear that it was "originally an unwritten talk, spoken from notes . . . edited from the taped transcription".[55] It emphasizes what she elsewhere stresses as the "elicited" nature of what she has written—a point in keeping with her concern with the problematical nature of writing for so many women—the difficulty not only of finding the time to write, the confidence to write, and the voice with which to write, but also of finding an audience for whom to write. It is, once again, a version of Olsen's basic stance of humility: she did not set out to write this essay, she was fortunate enough to be asked to write it—and without that encouragement, perhaps, she suggests, it would never have existed. (A similar headnote introduces the second essay, which was "an unwritten talk, spoken from notes. . . .")[56]

The first essay expands on the points Olsen makes in her final prefatory page. She makes it clear that the silences of which she speaks are widespread, and are known by "the great in achievement" as well as by those who never managed to achieve. She discusses silences in great writers where "the creative working atrophied and died."[57] She quotes from these writers' journals to reveal their anxieties, frustrations, and difficulties with sustaining the creative energy that fueled their greatest work.

She then directs the reader's attention to other kinds of silences:

> Kin to these years-long silences are the hidden silences; work aborted, deferred, denied—hidden by the work which does come to

fruition. . . . Censorship silences. Deletions, omissions, abandon-
ment of the medium . . . ; paralyzing of capacity. . . . Publishers'
censorship, refusing subject matter or treatment as "not suitable"
or "no market for." Self-censorship. Religious, political censorship. . . . Other silences. The truly memorable poem, story or
book, then the writer ceasing to be published. Was one work all the
writers had in them . . . and the respect for literature too great to
repeat themselves? . . . Were the conditions not present for establishing the habits of creativity? . . . —or—as instances over and
over—other claims, other responsibilities so writing could not be
first. (The writer of a class, sex, color still marginal in literature,
and whose coming to a written voice at all against complex odds is
exhausting achievement.)[58]

Olsen's list of silences goes on: "the foreground silences, *before*
the achievement," "the silences where the lives never came to
writing."[59] At this point Olsen explores, through their journals,
letters and notes, how various practitioners—people whose lives
did "come to writing"—explain "the work of creation and the circumstances it demands for full functioning."[60] After several pages
of extracts printed conventionally in paragraphs indented from her
own prose, Olsen quotes two pages of excerpts from Kafka's journal, each entry standing alone, framed by double-spaced blankness. It is a sign of things to come.

The remainder of Part One, essays titled "One Out of Twelve:
Writers Who Are Women in Our Century" and "Rebecca Harding
Davis," is, from a formal standpoint, fairly straightforward.[61] Part
Two, however, which is titled "Acerbs, Asides, Amulets, Exhumations, Sources, Deepenings, Roundings, Expansions,"[62] is far from
conventional. Indeed, the reader hardly knows what to make of it.
Olsen says, on the title page of Part Two, "Much of this af-
tersection is the words of others—some of them unknown or little
known, others of them great and famous. Each quotation, as each
reference to lives, is selectively chosen for maximum significance;
to become—or to again become—current; to occur and recur; to
aim. The organization follows the order of thought in the original
essays, page by page, and is so keyed."[63] This sounds suspiciously
like a section that, in more conventional books, would be labeled,
"Endnotes." Indeed, this is precisely where, in more conventional
books, endnotes would belong.

As if to confirm the reader's suspicions—and provide a context in which she can make sense of this mysterious "Acerbs, Asides, Amulets" section, Olsen follows the word "Expansions" in this section's title, and the last word on this introductory page by a superscript solid triangle. The triangle appears again at the bottom of Part Two's title page with an explanation: "Essay page number to which each refers."[64] It looks like endnotes, and acts like endnotes, but is it endnotes? And if it is, why has Olsen gone to so much trouble to depart from the expected? The answer, I suggest, cuts to the core of Olsen's central theme.

The essays represent a triumph over silence. But the sparks that engendered those essays—the flashes of insight "long in accumulation, garnered over fifty years, near a lifetime,"—were themselves sandwiched into periods of silence, work, childrearing, self-doubt, and so forth. To show the finished products but not to give insight into the process that shaped them would do violence to Olsen's larger enterprise in the book, sharing with her reader her own odyssey through the mysteries of insight, awareness, creativity, confidence, and productivity—and the lack thereof. How do you communicate the fragmentary, uneven, sporadic nature of consciousness when the mere presence of your words on the page—copyrighted, published, sold in stores—attests to your ability to transform those uneven fragments into sustained narrative? How do you share with your reader the nature of those insights that land in the mind and explode like hand grenades with shattering or exhilarating reverberations—without your own prose filtering out their raw sound, their blinding light? Finally, how can you address the problem of interruptions, a problem particularly common in women's lives? How can you let your reader know that everything that stops her from reading your book straight through (her life, her constant interruptions) is a point of connection between the two of you rather than an obstacle to be overcome, and a point of connection between her and the women of whom you write?

Olsen addresses all of these questions by making Part Two a montage of fragments pasted on white space. Having voiced her themes initially in the first half of the book, she reiterates them here (italicizing them if they are word-for-word repetitions), and expands on them, sharing with the reader the thought processes that yielded them. The "Acerbs, Asides, Amulets" section of the book accepts the interruptions that are part of readers' lives as a given, and glories in them: it makes no difference if it is read straight through or in bits and pieces. In fact, one might argue that it is meant to be read as it was written, in bits and pieces, in time stolen between chores, allowing time to meditate on the significance of one passage before moving on to the next. In the blank spaces between the fragments, the reader is given permission to pause, to think, to insert her own response, to recall her own experience, to listen to her own voice.

Olsen ends the book not with her experimental "Acerbs, Asides, Amulets" section, but with excerpts from Rebecca Harding Davis's *Life in the Iron Mills*. The move makes sense, for while Olsen wants to encourage her reader to hear her own voice, a central theme throughout the book is the notion that by listening to the voices of others—particularly others who struggled to overcome their own silences and were then silenced by the world—one can create a space in which one's own voice, as well as those of others, may be heard.

GLORIA ANZALDÚA

The universe of discourse Gloria Anzaldúa's 1987 work *Border-lands/La Frontera* entered was, to some extent, one that she had helped to shape in key ways. An earlier text, the award-winning anthology *This Bridge Called My Back: Writing by Radical Women of Color*, edited by Anzaldúa and Cherríe Moraga, appeared in 1981, was reissued in 1983, and sold more than 35,000 copies. It also helped pave the way for Anzaldúa's virtuoso 1987 experiment.[65] Anzaldúa's contributions to the earlier book—her introduction, her essay, "Speaking in Tongues: A Letter to Third World Women Writers," her story, "La Prieta," her interview with Luisah Teish, and her foreword to the second edition—show some signs of things to come. There were other precursors and influences as well.

Previous nonfiction writing on Chicano life and on the pre-Columbian past (including journalism, history, sociology, and anthropology by both Anglos and Chicanos), as well as poetry and fiction by Chicanos, provide a context in which *Borderlands* might be read. Any reader who tries to fit *Borderlands* neatly into this context, however, will have trouble for Anzaldúa transforms, enlarges, deconstructs, and reconstructs the generic conventions out of which her own work grows.

Anzaldúa cites a number of works in both of these categories in the book itself, both in the text and in endnotes. The nonfiction texts she cites include books by journalists, historians, sociologists, anthropologists, and linguists, on such topics as the history of the Chicanos of Aztlán, the gods of ancient Mexico, Chicano images of the southwest, Anglo attitudes toward Mexicans in Texas, and the dialect of the barrio. Imaginative writing Anzaldúa cites in the book include *City of Night* by John Rechy and *I am Joaquín/ Yo Soy Joaquín* by Rodolpho Gonzales.[66] A number of women writers—such as Elena Poniatowska, Rosario Castellano, and Estela Portillo—also shaped Anzaldúa's thinking about her craft, although she does not cite them in the book.[67]

The nonfiction discussions of Chicano life and of the pre-Columbian past addressed in a clear, dry, scientific way subjects that held, for Anzaldúa, pain, exhilaration, irony, agony, unspeakable ugliness and luminous beauty—in short, the stuff of poetry and fiction. While they covered the territory, these texts failed to encompass not only the complex emotions involved, but the challenge—both the difficulty and the triumph—of dealing with these emotions in two, three, four—sometimes eight—different interweaving languages.[68]

Some of the poetry and fiction, on the other hand, did blend Spanish and English; Anzaldúa recalls her excitement at finding, in *I am Joaquín/ Yo Soy Joaquín* (1967) a truly bilingual book by a Chicano.[69] But the poets' and novelists' world was the world of the imaginative artist, well-stocked with both truth and beauty. It was not endowed, however, with much authority for the accuracy of its reports in the world outside the world the artist made.

In *Borderlands/ La Frontera,* Gloria Anzaldúa merges two narrative realms: that of the poet and that of the nonfiction writer. While moving between Spanish and English (in addition to Nahuatl and a range of other languages) with the skill of the poet, Anzaldúa lays out the facts of her subject with the care of the historian or social scientist. The result is a book unlike any that had gone before.[70] It is Anzaldúa's artful movement between Spanish and English that makes the book a groundbreaking experiment in American nonfiction narrative, and it is the dynamics of that movement that will concern us here.

The reader who wants to understand *Borderlands/La Frontera*, must begin, as the reader of *Silences* must begin, with the acknowledgments. "To you who walked with me upon my path and who held out a hand when I stumbled," Anzaldúa writes,

> to you who brushed past me at crossroads never to touch me again;
>
> to you whom I never chanced to meet but who inhabit borderlands similar to mine;
>
> to you for whom the borderlands is unknown territory;[71]

Addressing, as she does, those who helped her, and those who brushed by, those who live in borderlands similar to hers, and those to whom borderlands are unknown, Anzaldúa manages to address her book, in the end, to everyone. Next comes a list of specific names of individuals, including Anglo names and Spanish names, a sign of things to come. It is followed by the first Spanish phrase to appear in the book, an italicized phrase, colloquial in its affectionate use of the diminutive, a simple "thank you" to all who helped: "*gracias a toditos ustedes.*" The Acknowledgments closes with the statement, "THIS BOOK is dedicated *a todos mexicanos* on both sides of the border." With a simple statement of dedication, Anzaldúa flags the subject of her book, and broaches the nature of the border as a line that divides "*mexicanos*" from "*mexicanos,*" something imposed, not "natural," something artificial, divisive, complex.

The preface that follows explains more precisely the central subject of the book. "The actual physical borderland that I'm dealing with in this book is the Texas–U.S. Southwest/Mexican border," Anzaldúa writes. But, she continues, "the psychological borderlands, the sexual borderlands and the spiritual borderlands are not particular to the Southwest. In fact, the Borderlands are physically present wherever two or more cultures edge each other, where people of different races occupy the same territory, where under, lower, middle and upper classes touch, where the space between individuals shrinks with intimacy."[72] In its ambitious efforts to link geographic, social historical, political, psychological, sexual, and spiritual issues, this book might be thought of as having more in common with Walt Whitman's 1855 *Leaves of Grass* than with any of its more immediate precursors: her concern is not only with the relationships between individuals and between communities, but with the complex, fragmented nature of subjectivity itself. What stops this far-ranging book from exploding with centrifugal force, however, is Anzaldúa's transformation of the linguistic borderland between Spanish and English into a metaphor for the nature of all the other borderlands.

The first time Anzaldúa mentions the state that borders Mexico she uses its English name, Texas; the second time she calls it by its Spanish name, *Tejas*. She is teaching readers Spanish as she goes along, easing them into a contextual awareness of what the words mean. "However, there have been compensations for this *mestiza,* and certain joys."[73] Anzaldúa's use of the Spanish word *mestiza* comes only after a contextual definition of it has been made available to the reader: "Mexican (with a heavy Indian influence) and the Anglo." The shift between English and Spanish has been very slight and subtle up to this point. Before she throws full-blown phrases in Spanish at the reader (*"la madre naturaleza"*) or words in the Aztec language, Nahuatl ("the wind, *Ehecatl"*), Anzaldúa reiterates both the problem that is the focus of her concern and the solution that animates her project as an artist: "Living on borders and in margins, keeping intact one's shifting and multiple identity and integrity, is like trying to swim in a new element, an 'alien' element. There is an exhilaration in being a participant in the future evolution of human kind."[74]

Anzaldúa then explains the switching of linguistic codes that will take place in the book:

> The switching of "codes" in this book from English to Castillian Spanish to the North Mexican dialect to Tex-Mex to a sprinkling of Nahuatl to a mixture of all these, reflects my language, a new language—the language of the Borderlands. There, at the juncture of cultures, languages cross-pollinate and are revitalized; they die and are born. Presently this infant language, this bastard language, Chicano Spanish, is not approved by any society. But we Chicanos no longer feel that we need to beg entrance, that we need always to make the first overture—to translate to Anglos, Mexicans and Lati-

nos, apology blurting out of our mouth with every step. Today we
ask to be met halfway. This book is our invitation to you—from the
new mestizas.[75]

She will not translate at every turn, or apologize with every step:
sometimes she will translate directly, sometimes she will paraphrase, and sometimes she will refuse to translate at all. The final
word of Anzaldúa's preface, "mestizas," is unitalicized: it is English. The lack of italics reminds the reader of the cross-pollinization
that has already taken place between English and Spanish, and
reminds us that we already inhabit a place where "two or more
cultures edge each other."
     The borderlands Anzaldúa charts in *Borderlands/La Frontera*
are legion: that between Chicano and Anglo, Mexican and Anglo,
Mexican and Chicana, Mexican and Chicano, Chicano and Chicana, Chicana and Chicana, Anglo and Indian, Indian and Mestizo, Spanish and Indian, Spanish and Mestizo, male and female,
female and female, male and male, heterosexual and homosexual,
rich and poor, urban and rural, spirit-world and real world, articulate and inarticulate, and so on. All of us, as we soon realize,
inhabit many borderlands. Anzaldúa invites us to enter hers, in
empathy and imagination; and she invites those who already inhabit that borderland to understand, appreciate, and reclaim it
with the same passionate concentration that yielded her art. "I am
a border woman," Anzaldúa writes. "I grew up between two cultures, the Mexican (with a heavy Indian influence) and the Anglo
(as a member of a colonized people in our own territory). I have
been straddling that *tejas*–Mexican border, and others, all my life.
It's not a comfortable territory to live in, this place of contradictions. Hatred, anger and exploitation are the prominent features of
this landscape."[76]
     Anzaldúa's book is comprised not only of multiple linguistic
codes, but also of multiple genres. Her first chapter, a powerful
essay embracing autobiography, history, journalism, and social science, starts with two extracts and a poem. The first extract, in
Spanish, is a quote from *"Los Tigres Del Norte,"* which the foot-
note identifies as a *"conjunto* band." The second extract is from a
book. It identifies *"The Aztecas del norte"* as "the largest single
tribe or nation of Anishinabeg (Indians) found in the United States

today. . . . Some call themselves Chicanos and see themselves as people whose true homeland is *Aztlán* [the U.S. Southwest]." We move from self-identification of *"el otro Mexico"* by the *conjunto* to a scholar's discussion of Chicanos' identification of themselves and their homeland to Anzaldúa's own statement on the subject— the first harsh and lyrical poem of many such poems the reader will encounter in the volume, one filled with striking images:

> 1,940 mile-long open wound
> > dividing a *pueblo*, a culture,
> > running down the length of my body,
> > > staking fence rods in my flesh,
> > > splits me   splits me

Even before Anzaldúa begins the narrative prose that comprises the bulk of this chapter, she has set the stage for the idea that borders are complex, elusive constructions that may bring to mind gashes, cuts, and deep painful wounds—images that are stark, fresh, and surprising.

For Anzaldúa,

> The U.S.–Mexican border *es una herida abierta* where the Third World grates against the first and bleeds. And before a scab forms it hemorrhages again, the lifeblood of the two worlds merging to form a third country—a border culture. Borders are set up to define the places that are safe and unsafe, to distinguish *us* from *them*. A border is a dividing line, a narrow strip along a steep edge. A borderland is a vague and undetermined place created by the emotional residue of an unnatural boundary. It is in a constant state of transition. The prohibited and forbidden are its inhabitants.[77]

Like Tillie Olsen, who is concerned only with unnatural silences, Anzaldúa is concerned only with unnatural boundaries (see Figure 6.4).

While the border between the United States and Mexico may seem "natural" from the standpoint of a culture that embraces the dominant version of U.S. political and military history, from the perspective of the descendant of the ancient tribes that lived throughout the region the border is an artificial construct, dividing a people from itself. Anzaldúa establishes the arbitrary nature of that construct, first by making clear that it was the Chicanos' ancestors who first occupied and ruled the land now known as the south-

western United States. "The oldest evidence of humankind in the
U.S.—the Chicano's ancient Indian ancestors—was found in Texas
and has been dated to 35,000 B.C. In the Southwest United States
archeologists have found 20,000-year-old campsites of the Indians
who migrated through, or permanently occupied, the Southwest,
Aztlán—land of the herons, land of whiteness, the Edenic place of
origin of the Azteca."[78] The unnatural quality of the border is
further established by Anzaldúa's dramatic autobiographical account, on the same page, of what happened to her cousin:

> In the fields, *la migra.* My aunt saying, *"No corran,* don't run.
> They'll think you're *del otro lao."* In the confusion, Pedro ran,
> terrified of being caught. He couldn't speak English, couldn't tell
> them he was fifth generation American. *Sin papeles*—he did not
> carry his birth certificate to work in the fields. *La migra* took him
> away while we watched. *Se lo llevaron.* . . . I saw the terrible weight
> of shame hunch his shoulders. They deported him to Guadalajara
> by plane. The furthest he'd ever been to Mexico was Reynosa, a
> small border town opposite Hidalgo, Texas, not far from McAllen.
> Pedro walked all the way to the Valley. *Se lo llevaron sin un centavo
> al pobre. Se vino andando desde Guadalajara.*[79]

One of Anzaldúa's key themes is the difficulty of translating
your life into a language that cannot express it. While she wants to
communicate the difficulties of living on linguistic borderlands, she
also wants to communicate the richness it entails—not only for
those who live there already, but, potentially, for the rest of us. She
pursues these goals with two complementary strategies. On the
one hand, the book is filled with highly evocative, clear, and readable passages—like the preceding story—that move back and forth
between Spanish and English in an apparently seamless flow, mini-
mizing the difficulty of translating from one realm to the other;
these passages give the reader a sense of what it feels like to live in
two languages, in a world where the borders between languages
are fluid. On the other hand, the book is also filled with slyly
evasive "nontranslations"—passages that might appear, on the surface, to be translations, but which turn out, on closer view, to be
nothing of the sort—complex passages that paraphrase, amplify,
contradict, extend, or contract the passages that precede them,
doing everything, in fact, but "translating" them. These passages

underscore for the reader the poverty of one who does not have a mestiza consciousness of language. Anzaldúa's goal, of course, is to teach the reader to appreciate and value precisely the fecundity of imagination and expression that *"la consciencia de la mestiza"* can yield. The book is interesting and readable on any terms, but for the reader who can approach the mestiza's linguistic awareness, it holds special secret luminous riches.

Many of the "approximate" translations occur when Anzaldúa is quoting colloquial speakers: they would never make in English the comments they make in Spanish, so Anzaldúa translates only the gist of what they say, paraphrasing loosely. Thus she writes "Our mothers taught us well, '*Los hombres nomás quieren una cosa*' men aren't to be trusted, they are selfish and are like children."[80] The reader is motivated to make sense of the twisting maze of Spanish and English in the text by her knowledge of what is left out if she reads in English alone, for the language spoken by "our mothers" and "the culture" is Spanish: "Through our mothers, the culture gave us mixed messages: *No voy a dejar que ningún pelado desgraciado maltrate a mis hijos.* And in the next breath it would say, *La mujer tiene que hacer lo que le diga el hombre.* Which was it to be—strong, or submissive, rebellious or conforming?"[81] Spanish is also the language in which folk traditions in the family are described. When they are translated into English they become more abstract and less concrete; their particulars live in Spanish alone:

> *Mi mamagrande Ramona toda su vida mantuvo un altar pequeño en la esquina del comedor. Siempre tenía las velas prendidas. Allí hacía promesas a la Virgen de Guadalupe.* My family, like most Chicanos, did not practice Roman Catholicism, but a folk Catholicism with many pagan elements.[82]

If Spanish is the language of a mother's admonitions to her daughter and of the folk traditions in the home, it is also the language of Anzaldúa's own introspection and solitude. Translating these moments directly would do violence to them, since the language in which they are experienced has a great deal to do with what is experienced:

> *Internada en mi cuarto con mi intocada piel, en el oscuro velo con la noche. Embrazada en pesadillas, escarbando el hueso de la ternura me envejezco. Ya verás, tan bajo que me he caído.*[83]

The paraphrase that follows makes no attempt to translate directly; rather, it expands and amplifies the text that precedes it: "I locked the door, kept the world out; I vegetated, hibernated, remained in stasis, idled. No telephone, no television, no radio. Alone with the presence in the room. Who? Me, my psyche, the Shadow-Beast?"[84] For Anzaldúa the writer, two languages are infinitely richer than one.

Anzaldúa wants to do more than simply validate her language or her vision: she wants to reclaim and celebrate dimensions of the experience they embody. Anzaldúa will settle for nothing less than wrenching from her own experience as an outsider—as a Chicana, as a mestiza, and as a lesbian—the guideposts of a new way of being in the world. Anzaldúa universalizes the particulars of her own experience by forging a powerful link between those particulars and the creative process itself. She makes a strong case for the idea that the sense of disruption, discontinuity, dislocation, ambiguity, uncertainty, and fear that living on cultural, social, linguistic, and sexual borders entails can be a positive force for the artist. For while living with constant fear (of humiliation, of deportation, of hundreds of unnamed terrors) can paralyze, it can also produce a heightened sensitivity to one's environment, or what Anzaldúa calls "*la facultad.*" Born of the fear that comes from paying intense attention to "when the next person is going to slap us or lock us away," this supersensitive "radar" that yields subtle and complex data about emotions and people gives "the one possessing this sensitivity" the sense of being "excruciatingly alive to the world."[85] "Another aspect of this faculty," Anzaldúa writes, is that "it is anything that breaks into one's everyday mode of perception, that causes a break in one's defenses and resistance, anything that takes one from one's habitual grounding, causes the depths to open up, causes a shift in perception. . . ." And while the disruption, dislocation, and ambiguity produce enormous tensions, the individual who learns to manage those tensions develops extraordinary capacities to forge new creative syntheses, to thrive in what Anzaldúa calls "the *Coatlicue* state." "*Coatlicue,*" a concept drawn from early Mesoamerican fertility myths,[86] "is a rupture in our everyday world . . . a synthesis of duality, and a third perspective—something more than mere duality or synthesis of duality."[87] Like Anzaldúa herself, "*Coatlicue*" is a symbol of the fusion of opposites.

Anzaldúa wants to share with the reader that sense of breaking through one's "everyday mode of perception." She wants to take her reader beyond her "habitual grounding"—whether that reader is a Chicana who is ashamed of her language, a lesbian who is afraid to express her love of women, an Anglo who may suspect the words "Chicano culture" to be an oxymoron, a poet who thinks history should be left to the historians, or a man who takes women's silence as tacit assent.

It is to this new awareness that the mestiza is peculiarly suited to lead us, Anzaldúa writes, in the section titled "El camino de la mestiza/The Mestiza Way."

> Just what did she inherit from her ancestors? This weight on her back—which is the baggage from the Indian mother, which the baggage from the Spanish father, which the baggage from the Anglo?
> *Pero es difícil* differentiating between *lo heredado, lo adquirido, lo impuesto.* She puts history through a sieve, winnows out the lies, looks at the forces that we as a race, as women, have been a part of . . . This step is a conscious rupture with all oppressive traditions of all cultures and religions. She communicates that rupture, documents the struggle. She reinterprets history and, using new symbols, she shapes new myths. She adopts new perspectives toward the darkskinned, women and queers. . . . She is willing to share, to make herself vulnerable to foreign ways of seeing and thinking. She surrenders all notions of safety, of the familiar. Deconstruct, construct . . .[88]

The new mestiza, Anzaldúa writes,

> copes by developing a tolerance for contradictions, a tolerance for ambiguity. . . . She learns to juggle cultures. She has a plural personality, she operates in a pluralistic mode. . . . Not only does she sustain contradictions, she turns the ambivalence into something else.[89]

The hundred pages of poems that follow the hundred pages of prose in the book serve as a reprise, chanting the themes the first section introduced in new and different keys, with subtle variations, all the while reaffirming their basic message. Anzaldúa reclaims ancient myth, recent history, contemporary conditions, and the timeless patterns of nature, and weaves them into a stunningly coherent paean to the anguish and beauty of life. "To survive the

Borderlands," Anzaldúa writes, "you must live *sin fronteras*/be a crossroads." It is to these crossroads that Gloria Anzaldúa's "new mestiza" will lead.

LEGACIES

Du Bois's *The Souls of Black Folk,* Agee's *Let Us Now Praise Famous Men,* Olsen's *Silences,* and Anzaldúa's *Borderlands/La Frontera* have more in common than the boldness of the formal experiments they embody. Each experiment was the result, at least in part, of the author's decision to expand the conventions of nonfiction narrative to include conventions drawn from another world entirely. And each experiment has served to inspire and engender an impressive literary legacy.

Du Bois's *The Souls of Black Folk* cast a net of influence that was both far-flung and deep. The critic Saunders Redding referred to the book as "more history-making than historical" in its profound impact on its time.[90] James Weldon Johnson wrote that it "had a greater effect upon and within the Negro race in America than any other single book published in this country since *Uncle Tom's Cabin.*"[91] For Du Bois, the spirit of another principally "oral tradition"—a folk culture that "high culture" artists such as Dunbar and Chesnutt were just beginning to explore in their creative work—would be central to his narrative in ways that it had never previously been to nonfiction. His experiment made many of the existing ways of writing about race in America seem lacking; it laid the groundwork for new departures in poetry and fiction, as well as nonfiction—a path that would never apologize for finding in the unwritten tradition of the "folk" the tropes that could pave the way for new understanding and new artistic triumph. In contemporary fiction, one thinks of Sherley Anne Williams's novel, *Dessa Rose,* in which refrains of "sorrow songs," not unlike those Du Bois recorded, frame a rich and powerful fictional narrative, or Gwendolyn Brooks's poem, "Queen of the Blues," in which song lyrics form a gloss on the singer's life.[92] In nonfiction, one sees signs that the project of linking folk culture with high culture is central to contemporary critical inquiry on race and culture in Houston Baker's work on the blues and African-American literature, in the exploration of the trope of the "signifying monkey" in African-American culture by Henry Louis Gates, Jr., or in Robert Stepto's use of the "call and response" as the organizing metaphor for African-American literary history.[93]

For Agee, the breakthrough would be tied to a close attention to material culture—to learning how to "read" the contours of a life in the decorations of its habitations, in the feel of its clothing, or in the taste of its food. His achievement made all previous writing that was less concrete, less attentive, and less reflective seem thin and wanting. It would help make possible the careful observations of Robert Coles, for example, or the attentive listening of Studs Terkel, both of whom acknowledge their debt to Agee. It would also help inspire the experiments of Tillie Olsen.

For Olsen, too, the key would be the contemplation of realms not usually "written down"—of the silent dance of tasks performed ritually and repeated endlessly by women throughout the ages. Olsen's probing of the nonwritten channels into which women's energies have been poured have helped spark a veritable renaissance of imaginative responses including poetry, fiction, drama, history, biography, journalism, criticism, oral history, and autobiography.[94]

Gloria Anzaldúa's creative exploration of mestiza consciousness and of linguistic borderlands has prompted other Chicanas to tell their own stories. It has also helped inspire writers who inhabit borderlands other than hers—in Puerto Rico, in Mexico, in Montreal—to pursue their own literary experiments linking language and identity.[95]

In each case, the writer gave the existing conventions of nonfiction narrative a heady infusion of traditions from another world: a world that had rarely if at all been recorded in print on its own terms. Each of these books—a triumph, as Du Bois put it, of "stalwart originality"—is a tour de force in part because of its commitment to giving form to the invisible, to making the unseen seen, to showing substance where only shadow was thought to be. In the process of bringing off this consummate performance, each book unlocked for future writers realms that, on the printed page, at least, had not even seemed to exist before.

Each of these books does more than bear witness to social and cultural realities that cannot and must not be ignored; each book bears witness to the power of the artist to change our perceptions by breaking through the familiar with boldly original imaginative strokes. Each of these transformative works of literary nonfiction helped clear the way for new kinds of writing about American culture—writing rooted in fact that is shot through with the poetry and passion of fiction, cultural reports bursting with the energy of felt life and with the power to convey important truths about that life. These books and a handful of others, such as John Hersey's *Hiroshima,* Penny Lernoux's *Cry of the People,* and Michael Harrington's *The Other America,* leave a legacy of conscience, clarity, responsible research, rhetorical power, fearless experimentation, and consummate artistic excellence—a towering standard against which literary nonfiction in the twenty-first century will undoubtedly be judged.

III

Of Mr. Booker T. Washington and Others

From birth till death enslaved; in word, in deed, unmanned!

Hereditary bondsmen! Know ye not
Who would be free themselves must strike the blow?

Byron.

Figure 6.1.

(But *I* am young; and I am young, and strong, and in good health; and I am young, and pretty to look at; and I am too young to worry; and so am I, for my mother is kind to me; and we run in the bright air like animals, and our bare feet like plants in the wholesome earth: the natural world is around us like a lake and a wide smile and we are growing: one by one we are becoming stronger, and one by one in the terrible emptiness and the leisure we shall burn and tremble and shake with lust, and one by one we shall loosen ourselves from this place, and shall be married, and it will be different from what we see, for we will be happy and love each other, and keep the house clean, and a good garden, and buy a cultivator, and use a high grade of fertilizer, and we will know how to do things right; it will be very different:) (? :)

((?)) :)

How were we caught?

What, what is it has happened? What is it has been happening that we are living the way we are?

The children are not the way it seemed they might be:

She is no longer beautiful:

He no longer cares for me, he just takes me when he wants me:

There's so much work it seems like you never see the end of it:

I'm so hot when I get through cooking a meal it's more than I can do to sit down to it and eat it:

How was it we were caught?

Figure 6.2.

*Blight never does good to a tree . . . but if it
still bear fruit, let none say that the fruit was
in consequence of the blight.*

—William Blake

. . . And yet the tree did bear fruit.

BLIGHT▲

Its Earliest Expression (Early 1600s)

I never rested on the Muses bed,
Nor dipt my quill in the Thessalian fountaine,
My rustick Muse was rudely fostered
And flies too low to reach the double mountaine.

Then do not sparkes with your bright Suns compare,
Perfection in a Woman's work is rare.
From an untroubled mind should verses flow,
My discontents make mine too muddy show.

And hoarse encumbrances of householde care,
Where these remain, the Muses ne'er repaire.

—Mary Oxlie of Morpet to
William Drummond of Hawthornden

*Compared to men writers of like distinction and years of
life, few women writers have had lives of unbroken productivity, or leave behind a "body of work." Early beginnings, then silence; or clogged late ones (foreground silences); long periods between books (hidden silences);
characterize most of us.*

▲ ONE OUT OF TWELVE, P. 38

Figure 6.3.

Tension grips the inhabitants
of the borderlands like a virus. Ambivalence and unrest reside
there and death is no stranger.

In the fields, *la migra*. My aunt saying, "*No corran*, don't run. They'll think you're *del otro lao*." In the confusion, Pedro ran, terrified of being caught. He couldn't speak English, couldn't tell them he was fifth generation American. *Sin papeles*—he did not carry his birth certificate to work in the fields. *La migra* took him away while we watched. *Se lo llevaron*. He tried to smile when he looked back at us, to raise his fist. But I saw the shame pushing his head down, I saw the terrible weight of shame hunch his shoulders. They deported him to Guadalajara by plane. The furthest he'd ever been to Mexico was Reynosa, a small border town opposite Hidalgo, Texas, not far from McAllen. Pedro walked all the way to the Valley. *Se lo llevaron sin un centavo al pobre. Se vino andando desde Guadalajara.*

Figure 6.4.

NOTES

Shelley Fisher Fishkin would like to thank the American
Council of Learned Societies for support that helped make
the writing of this chapter possible.

1. See, for example, letters from Columbus, Verrezano, and Cabeza de Vaca in *The Harper Anthology of American Literature,* Vol. 1, Donald McQuade, ed. (N.Y.: Harper and Row, 1987), pp. 10–36. See also Thomas Jefferson, "Notes on the State of Virginia," in *The Portable Thomas Jefferson,* Merrill D. Peterson, ed. (New York: Viking, 1975), pp.187–88.

2. I. Garland Penn, *The Afro-American Press and Its Editors,* originally published 1891; Reprint edition (Salem, New Hampshire: Ayer Company, 1988), p.114.

3. Arnold Rampersad, *The Art and Imagination of W. E. B. Du Bois* (Cambridge: Harvard University Press, 1976), p. 10. (The *Globe* later became the *Freeman* and finally the *Age.*)

4. ibid., p. 21.

5. ibid., pp. 515–17.

6. See for example: "Strivings of the Negro People," *Atlantic Monthly* 80: 194–98 (August 1897), which would form the basis of Chapter 1 of *Souls;* "The Freedman's Bureau," *Atlantic Monthly* 87:354–65 (March 1901), which would become Chapter 2; "The Evolution of Negro Leadership," *Dial* 31: 53–55 (July 16, 1901) (Chapter 3); "A Negro Schoolmaster in the New South," *Atlantic Monthly* 83: 99–104 (January 1899) (Chapter 4); "Of the Training of Black Men," *Atlantic Monthly* 90: 289–97 (September 1902) (Chapter 6). For additional journalism that fed into *Souls,* see Rampersad, *The Art and Imagination of W. E. B. Du Bois,* p. 303. In 1910 Du Bois would change the shape of African-American journalism in America by founding *Crisis,* the influential NAACP-backed publication that he would edit and contribute to for many years.

7. Robert Stepto has noted that "Du Bois's jouney in *The Souls* from infancy in the Berkshires of western Massachusetts to adulthood amid the western hills of Atlanta, complete with symbolic disembarkings in Tennessee, Philadelphia, and that part of the Black Belt surrounding Albany, Georgia, is not simply the story line of the volume but also the narrative manifestation of Du Bois's cultural immersion ritual. What is extraordinary and absolutely fresh about this ritual is that, in terms of its symbolic geography, it is a journey both to and into the South." Stepto also observes that in general, "prior to *The Souls,* the seminal journey in Afro-American narrative literature is unquestionably the journey north." Robert Stepto, *From Behind the Veil: A Study of Afro-American Narrative* (Urbana: University of Illinois Press, 1979), pp. 66–67.

8. Slave narratives: Frederick Douglass, *Narrative of the Life of Frederick Douglass, an American Slave, Written by Himself* (1845); Henry Bibb, *Narrative of the Life and Adventures of Henry Bibb, an American Slave* (1849); William Wells

Brown, *Narrative of the Life and Escape of William Wells Brown* (1852), Solomon Northrup, *Twelve Years a Slave* (1854); and Harriet Jacobs, *Incidents in the Life of a Slave Girl* (1861).

Histories: William Stills, *The Underground Railroad* (1872); George Washington Williams, *History of the Negro Race in America from 1619 to 1880: Negroes as Slaves, Soldiers and as Citizens,* 2 vols. (1882–1883); 2 vols in one (1885); also Williams, *A History of the Negro Troops in the War of the Rebellion, 1861–1865* (1887); Joseph T. Wilson, *The Black Phalanx: A History of the Negro Soldiers of the United States in the Wars of 1775–1812, 1861–1865* (1887).

9. Anna Julia Cooper, *A Voice From The South* (1892); Ida Wells-Barnett, *A Red Record* (1895).

10. W. E. B. Du Bois, *Dusk of Dawn,* in *W.E.B. Du Bois: Writings,* Nathan Huggins, ed. (New York: The Library of America, 1986), p. 596.

11. ibid., p. 590.

12. Rampersad, *The Art and Imagination of W. E. B. Du Bois,* p. 52

13. Du Bois, *Dusk of Dawn,* p. 599.

14. ibid., p. 591.

15. ibid., pp. 602–3.

16. Du Bois also came to feel that "there was no such definite demand for scientific work of the sort that I was doing, as I had confidently assumed would be easily forthcoming." *Dusk of Dawn,* p. 603.

17. While several parts of *Souls* were published in periodicals in the late 1890s, it is the final book-length manuscript that interests us here. Detailed comparisons of some of Du Bois's revisions of earlier magazine pieces into several of the chapters of *Souls* (an analysis that does not involve the points raised here) may be found in Robert Stepto's *From Behind the Veil: A Study of Afro-American Narrative,* pp. 52–63.

18. W. E. B. Du Bois, *The Souls of Black Folk* , Library of America edition of writings cited earlier, pp. 359–60.

19. ibid., pp. 363, 372, 392, 405, 415, 424, 439, 456, 475, 493, 506, 512, 521.

20. ibid., p. 405, translation p. 1321 n.

21. ibid., pp. 536–37.

22. ibid., p. 538.

23. ibid., p. 538.

24. ibid., p. 539.

25. ibid., p. 544.

26. ibid., pp. 544–45.

27. W. E. B. Du Bois, "The Conservation of Races," American Negro Academy Occasional Papers, no. 2 (Washington, D.C., 1897), quoted in Rampersad, *op. cit.,* p. 61.

28. William Stott, *Documentary Expression and Thirties America* (New York: Oxford University Press, 1973), p. 261.

29. Genevieve Moreau, *The Restless Journey of James Agee* (New York: William Morrow and Co., Inc., 1977), p. 133.

30. Stott, *Documentary Expression and Thirties America,* p. 211.

31. ibid., p. 211.

32. ibid., p. 212.

33. ibid, p. 221..

34. ibid., p. 222.

35. ibid., p. 223.

36. Victoria A. Kramer, *James Agee* (Boston: G.K. Hall, 1975), p. 75.

37. Published in 1947 by Houghton Mifflin, only 1,025 copies were sold before the book went out of print in 1948. Herbert Mitgang, "50 Years After Portrait, Agee's Poor Still Are," *The New York Times* (Books of the Times), June 17, 1989, National Edition, p. 13.

38. Agee letter quoted in *Agee: His Life Remembered*, Ross Speares and Jude Cassidy, eds. (New York: Holt, Rinehart and Winston, 1985), p. 75.

39. Moreau, p. 165.

40. ibid., p. 166.

41. James Agee and Walker Evans, *Let Us Now Praise Famous Men* (New York: Ballantine Books, 1966), p. xiv.

42. ibid., p. 12.

43. ibid., p. xiv.

44. ibid., p. xv.

45. ibid., pp. 71–77; p. 405.

46. Stott, *Documentary Expression and Thirties America*, p. 304.

47. Not withstanding Agee's diffidence and professions of humility, some of the people about whom he wrote came to resent deeply what they came to think of as Agee's exploitative distortions of their lives. For example, one man expressed his candid resentment in front of the camera—a video camera this time around—in the film "*Let Us Now Praise Famous Men* Revisited," a production in "The American Experience" series, WGBH-TV Boston, first telecast November 29, 1988. This film is a revision of "Alabama: 40 Years On," directed by Carol Bell, produced by BBC-TV, 1979. Others expressed their resentment to journalists Dale Maharidge and Michael Williamson who interviewed 128 survivors and descendents of the twenty-two people Agee had written about originally, in *And Their Children After Them* (New York: Pantheon, 1989).

48. Tillie Olsen, *Silences* (1978) (New York: Delta/Seymour Lawrence paperback, 1989). For more on the biographical backgrounds of Olsen's work, see Selma Burkom and Margaret Williams, "Deriddling Tillie Olsen's Writings," *San Jose Studies* 2: 65–83 (1976), and Deborah Rosenfelt, "From the Thirties: Tillie Olsen and the Radical Tradition," in *Feminist Criticism and Social Change*, (New York: Methuen, 1985), pp. 216–48.

49. Tillie Olsen, personal communication.

50. Tillie Olsen, "Silences: When Writers Don't Write," *Harper's Magazine* 231: 153–61 (October 1965).

51. Writers who had publicly acknowledged the importance of Olsen's work on the theme of "silences" by 1978 include Adrienne Rich and Susan Griffin. Later acknowledgments came from such critics as Lillian Robinson, Jane Marcus, Florence Howe, Annette Kolodny, Deborah Rosenfelt, Elaine Hedges, Catherine Stimpson, Bonnie Zimmerman, Paul Lauter, Alice Walker, Anne Sexton, Rosario Ferre, Susan Griffin, Hortense Spillers, Margaret Randall, Elizabeth Meese, Lee

Edwards, Arlyn Diamond, Linda Wagner-Martin, and many others. (See note 94).

52. Olsen's "Acknowledgment" pages are worth attention in their own right. The reader who has completed *Silences* and then returns to its beginning will recognize that the experimental fragments of "Part Two" are, in fact, echoes of what the reader receives on the book's earliest pages. When these pages are confronted initially, however, they appear neither experimental nor strange: they are merely the stuff with which so many books begin—acknowledgments, dedication, prefatory remarks, table of contents. It is a testimony to Olsen's art, and to her ability to transform indifferent or hostile conventions into useful tools that these "standard" features take on, by the end of her book, special resonance. Acknowledgments are generally routine affairs. But for Tillie Olsen, as the reader will eventually come to see, acknowledgments—and copyright notices and the lack thereof—are, in a sense, as much the subject of her book as anything else. These two factors: the existence of texts, or the lack of texts, and the existence of conditions (individuals, institutions, funding, encouragement) supportive of creative work, or the lack thereof, are precisely what this book is about. The space devoted to them serves to alert the reader early on that the personal—the circumstances that led *this* writer to write *this* book—is more than purely personal; it is political and it is germane. And it is to be foregrounded early and often.

53. These pages are not numbered. They can be described only by their positiion in the text of *Silences*.

54. Olsen, *Silences,* (ix).

55. Olsen, *Silences,* p. 5.

56. ibid., p. 22.

57. ibid., p. 6.

58. ibid., pp. 8–9.

59. ibid., p. 10.

60. ibid., p. 11.

61. Olsen does experiment somewhat with typography and layout in the second essay, "One Out of Twelve: Writers Who Are Women in Our Century," in which she explores the problem of the disproportionate distribution of silences among women. Pages in this chapter occasionally resemble in layout those quoted from Kafka's journal not because they are journal extracts, but because Olsen weaves back and forth between quotes from women's texts, her own narrative, and the amplifying or ancillary ideas that occur to her as she writes—or after she has written?—but which do not belong in the body of the text; this later category of material appears as asterisked notes at the bottom of the page. They include such intriguing items as Olsen's list of recent writers with children, or her list of those rare women writers whose work was encouraged, elicited, and facilitated by another woman serving as the "essential angel"—a role women often play for male writers, but all too rarely, Olsen suggests, for other women. Olsen's decision to subordinate points like these, and to separate them from the body of her main text, helps emphasize that they are exceptions that prove the rule. Her main concern is with the fact that *most* women writers in our century were, in fact, childless, and most did *not* have a facilitating angel to act as midwife to their work.

62. Olsen, *Silences,* pp. 119–283.

63. ibid., p. 119.

64. ibid., p. 119.

65. *This Bridge Called My Back: Writing by Radical Women of Color,* Cherríe Moraga and Gloria Anzaldúa, eds. (New York: Kitchen Table: Women of Color Press, 1983). Anzaldúa contributed to two other anthologies of writing by Latinas, one published in 1983 and the other the same year as *Borderlands/La Frontera,* 1987. These two books also help provide a context for *Borderlands/La Frontera.* The anthologies are: Alma Gomez, Cherríe Moraga, and Mariana Romo-Carmona, eds., *Cuentos: Stories by Latinas* (New York: Kitchen Table: Women of Color Press, 1983) and Juanita Ramos, ed., *Compañeras: Latina Lesbians* (*An Anthology*), (New York: Latina Lesbian History Project, 1987). Also relevant is *Third Woman Journal,* a publication edited by Norma Alarcón that has expanded on subjects broached in *This Bridge Called My Back* in a range of special issues during the last few years. (These include: *Third Woman Journal,* Vol. I, "Looking East (The Puerto Rican Issue)," Vol. II, "Hispanic Women: International Perspectives," Vol. III, "Texas and More," Vol. IV, "The Sexuality of Latinas.")

66. Jack D. Forbes, *Aztecas del Norte: The Chicanos of Aztlán* (Greenwich, Conn.: Fawcett Publications, Premier Books, 1973); John R. Chávez, *The Lost Land: The Chicano Images of the Southwest* (Albuquerque: University of New Mexico Press, 1984); Arnoldo De León, *They Called Them Greasers: Anglo Attitudes Toward Mexicans in Texas, 1821–1900* (Austin: University of Texas Press, 1983); C. A. Burland and Werner Forman, *Feathered Serpent and Smoking Mirror; The Gods and Cultures of Ancient Mexico* (New York: G.P. Putnam & Sons, 1975); R. C. Ortega, *Dialectologiá del Barrio,* Hortencia S. Alwan, trans. (Los Angeles: R. C. Ortega Publisher & Bookseller, 1977); Eduardo Hernández-Chávez, Andrew D. Cohen, and Anthony F. Beltramo, *El Lenguaje de los Chicanos: Regional and Social Characteristics of Language Used by Mexican Americans* (Arlington, VA: Center for Applied Linguistics, 1975); Rodolfo Gonzales, *I am Joaquín/ Yo Soy Joaquín* (New York: Bantam Books, 1972) (first published, 1967).

67. Personal communication, Gloria Anzaldúa, April 3, 1989. Anzaldúa noted that the portion of the book dealing with the influence of these women writers on her work had to be cut due to limits of space in the final edited version.

68. "How to Tame a Wild Tongue."

"Some of the languages we speak are:

1. Standard English
2. Working class and slang English
3. Standard Spanish
4. Standard Mexican Spanish
5. North Mexican Spanish dialect
6. Chicano Spanish (Texas, New Mexico, Arizona, and California have regional variations)
7. Tex-Mex
8. *Pachuco* (called *caló*)."

Anzaldúa, *Borderlands,* pp. 55–56.

69. ibid., p. 59.

70. Anzaldúa confirmed that, to her knowledge, the only books that moved back and forth between Spanish and English prior to hers were poetry or fiction. Her experiment was, quite self-consciously, an effort to transfer that technique to the realm of nonfiction. Personal communication. (The anthology to which Anzaldúa contributed and which appeared the same year *Borderlands* appeared, *Compañeras: Latina Lesbians*, is the only example that comes to mind of other writers' employing to any significant extent the kind of blending of Spanish and English that Anzaldúa pioneered. Several pieces of nonfiction in the book that move back and forth between Spanish and English include autobiographical narratives by Mirtha Quintinales ("Maybe You're Wrong, Papi") and Mariana Romo-Carmona ("*Una madre*"), and several interviews.

71. Anzaldúa, *Borderlands/La Frontera*, Acknowledgments (pp. i–ii)

72. ibid., Preface (v).

73. ibid.

74. ibid.

75. ibid., (vi).

76. ibid., (v).

77. ibid., p. 3.

78. ibid., p. 4.

79. ibid.

80. ibid., p. 17.

81. ibid., p. 18.

82. ibid., p. 27.

83. ibid., pp. 43–44.

84. ibid., p. 44.

85. ibid., p. 38.

86. ibid., p. 27

87. ibid., pp. 46–47.

88. ibid., p. 82.

89. ibid., p. 79.

90. Du Bois, *The Souls of Black Folk*, Saunders Redding, ed. (New York: Fawcett, 1961), p. ix.

91. James Weldon Johnson, *Along This Way* (New York: Viking, 1968), p. 203. Cited in Arnold Rampersad, *The Art and Imagination of W. E. B. Du Bois*, p. 68.

92. Sherley Anne Williams, *Dessa Rose* (New York: William Morrow, 1986) and Gwendolyn Brooks, "Queen of the Blues," reprinted in *The Norton Anthology of Literature by Women*, Sandra M. Gilbert and Susan Gubar, eds. (New York: Norton, 1985) p. 1858.

93. Houston Baker, *Blues, Ideology and Afro-American Literature: A Vernacular Theory* (Chicago: University of Chicago Press, 1984); Henry Louis Gates, Jr., *The Signifying Monkey;* and Robert Stepto, *From Behind the Veil: A Study of Afro-American Narrative* (Urbana: University of Illinois Press, 1979)

94. It helped inspire the fiction and poetry of Maxine Hong Kingston, Anne Sexton, Gloria Naylor, Sandra Cisneros, Ursula LeGuin, Helena Maria Viramontes, Alix Kates Schulman, Genny Lim, Margaret Laurence, Caryl Churchill, Margaret Atwood, Mary Stewart, Margaret Randall, Joyce Johnson, and Alice

Walker, as well as nonfiction by writers including Lillian Robinson, Jane Marcus, Florence Howe, Annette Kolodny, Deborah Rosenfelt, Elaine Hedges, Catherine Stimpson, Bonnie Zimmerman, Paul Lauter, Rosario Ferre, Susan Griffin, Hortense Spillers, Margaret Randall, Elizabeth Meese, Lee Edwards, Arlyn Diamond, Linda Wagner-Martin, and many others. For a detailed examination of the legacy *Silences* has left for critical and creative thinking, see Shelley Fisher Fishkin, "Reading, Writing and Arithmetic: The Lessons *Silences* Has Taught Us" (presented at the Modern Language Association, December, 1988. Forthcoming in *Listening to "Silences": New Essays in Feminist Criticism,* edited by Shelley Fisher Fishkin and Elaine Hedges.)

95. Personal communication, Gloria Anzaldúa. See, for example, *Cantando Bajito/Singing Softly* by Puerto Rican writer Carmen de Monteflores (San Francisco: Spinsters/Aunt Lute, 1989).

7

The Politics of the Plain Style

Hugh Kenner

Monsieur Jourdain, the Molière bourgeois, was so misguided as to conclude he'd been talking prose all his life, his bogus instructor having defined prose as whatever is not verse. But as nobody talks in rhyme, so nobody talks in prose. Prose came late into every written language, and very late into English. Even Chaucer had few clues to its workings. A special variety, "plain" prose, came especially late. Plain prose, the plain style, is the most disorienting form of discourse yet invented by man. Swift in the eighteenth century, George Orwell in the twentieth are two of its very few masters. And both were political writers—there's a connection.

The plain style has been hard to talk about except in circles. Can plainness, for instance, even lay claims to a style? Swift seems to think so. "Proper words in proper places" is what he has to say about style, not explaining, though, how to find the proper words or identify the proper places to put them into. But Swift is teasing. His readers in 1720 were among the first to feel alarm at the norm of the printed page, the way members of the intelligentsia in the 1950s were alarmed by television. Swift confronts them with their own bewilderment about what *style* may mean on silent paper, where words have not cadences or emphases but places. He is nearly asking if style has become a branch of geometry.

Styles were long distinguished by degrees of ornateness, the more highly figured being the most esteemed. There was a high style in which Cicero delivered his orations and a low style in which he would have addressed his cook. Rhetoricians gave their attention to the high style. The low style was beneath attention. It was scarcely, save by contrast, a style at all.

Evaluation like that has nothing to do with writing. It appeals to the way we judge oral performance. When Cicero spoke with his cook, he was offstage; when he addressed the Senate, he was in costume and in role and in command of a scene carefully prescripted. Of the five parts into which the Romans analyzed oratory, two pertained to the theatrics of performance; they were "memory" and "delivery." "Memory" is a clue to something important. Cicero's intricate syntax, its systems of subordination, its bold rearrangements of the natural order of words would have been impossible for an orator to improvise. So he worked them out on paper, memorized them and performed them in a way that made it seem he was giving voice to his passion of the moment. In fact, he was being careful not to let passion master him, lest it overwhelm memory.

A good public speech is something as contrived as a scene by Shakespeare. Even Lincoln, in what is represented as an address of exemplary plainness, launched it with diction he could only have premeditated: "Four score and seven years ago our fathers brought forth on this continent. . . ." The word *style* pertains to the art of contriving something like that. You contrive it by hand. A stilus was a pointed tool with which Romans wrote on wax tablets, and what you did with its aid was what came to be called your style. It was *you,* like your handwriting. A plain style would seem to be a contradiction in terms. If it's plain, then surely it didn't need working out with a stylus.

But indeed it did. Something so lucid, so seemingly natural, that we can only applaud its "proper words in proper places" is not the work of nature but of great contrivance. W. B. Yeats wrote, on a related theme:

> I said, "A line will take us hours maybe;
>
> Yet if it does not seem a moment's thought,
>
> Our stitching and unstitching has been naught."

Here's an intricate instance of writing that's saying it was spoken despite the fact that it rhymes, writing therefore that's inviting us to ponder its own degrees of artifice. Yeats has in mind poetry that has abandoned the high style and is managing to look not only improvised but conversational, even while rhyming. That would be poetry contriving, as T. S. Eliot put it, to be "at least as well written as prose." And it helps us perceive good prose as an art with a new set of norms—feigned casualness, hidden economy.

Since you're feigning those qualities, nothing stops you from feigning much more. George Orwell wrote "A Hanging," an eyewitness account of something he almost certainly never witnessed, as well as "Shooting an Elephant," his first-person recollection of a deed he may or may not ever have done.

We like to have such things plainly labeled fiction, if they are fictions. Then we are willing to admire the artistry—so acutely invented a detail as the condemned man's stepping around a puddle within yards of the rope, which prompts the narrator's reflection on "the unspeakable wrongness of cutting a life short when it is in full tide. This man was not dying, he was alive just as we were alive." That is like John Donne meditating on a sacred text, and we would not welcome news that the text was nowhere in the Bible, that Donne had invented it for the sake of the sermon he could spin. True, we can cite something Orwell wrote elsewhere, "I watched a man hanged once." Alas, that doesn't prove that he watched a man hanged once; it proves only that the author of "A Hanging" (1931) still had such an idea on his mind when he was writing something else six years later. An appeal to other writers may be more helpful. We soon find that Swift wrote a very similar sentence: "Last week I watched a woman *flay'd.*"

We could surely find more parallels, and in seeking them we'd be nudging "A Hanging" from reporting into literature, where questions of veracity can't reach it. For we'll half accept the idea that printed words do no more than permute other printed words, in an economy bounded by the page. That gets called the literary tradition, in which statements aren't required to be true.

But if we'll half accept the fictive quality of everything we read, don't we also tend to believe what it says in black and white, what we read in the papers? Of course we do, perhaps because the printed word stays around to be checked, like a stranger with nothing to hide. (Though handwriting does that too, print has the advantage of looking impersonal.)

Plain prose was invented among consumers of print, to exploit this ambiguous response. It seems to peg its words to what is persistently *so*—no matter how words drift about. Couched in plain prose, even the incredible can hope for belief. It's a perfect medium for hoaxes. By publishing the word that a nuisance, an astrologer named Partridge, was dead, Swift caused him vast trouble proving he was alive, and H. L. Mencken's mischievous printed statement that the first American bathtub got installed as recently as December 20, 1842, is enshrined as history in the *Congressional Record,* although Mencken himself tried to disavow it four times. Having grown famous for a baroque manner that advertised its own exaggerations, Mencken may have been surprised to find he could make people believe anything if he simply dropped to the plain style.

The science writer Martin Gardner, whose style is plain to the point of naïveté, had a similar experience when he sought to amuse readers of *Scientific American* by discussing a bogus force that haunted pyramids and could sharpen razor blades. The joke instantly got out of hand. Cultists of pyramid power made themselves heard, and Mr. Gardner has tried in vain to discredit them. Like Mr. Gardner's pyramid and Mencken's bathtub, the novel, which we both believe and don't, has origins inextricable from fakery. Readers in the eighteenth century could savor *The Life and Strange Surprising Adventures of Robinson Crusoe,* who'd been cast away on an island. That was an exotic thought if you lived in crowded London, and exoticism fostered the will to believe. The title page, moreover, said, "Written by Himself," so the account had the merit of firsthand truth. It was a while before "Himself" turned out to be a journalist named Defoe.

Today we handle the question of deception by saying Defoe was writing a novel, a genre of which he would have had no inkling. Defoe had simply discovered what plain prose, this new and seemingly styleless medium, is good for. Nothing beats it as a vehicle for profitable lies, which can entertain people and may even do them good in other ways. Knowing as we do that Defoe, not Crusoe, was the author, we still contrive to read *Robinson Crusoe* as if it were true. The formula "willing suspension of disbelief" was invented to help us accept what we are doing.

The next development was the replacement of the old polemical journalism by journalism of fact, meaning reports you could trust, statement by statement, fact by fact, because they appeared in newspapers. Gradually newspapers gravitated toward the plain style, the style of all styles that was patently trustworthy—in fact, the style of *Robinson Crusoe,* with which Defoe had invented such a look of honest verisimilitude. A man who doesn't make his language ornate cannot be deceiving us; so runs the hidden premise. Bishop Thomas Sprat extolled "a close, naked, natural way of speaking" in 1667; it was the speech, he went on to say, of merchants and artisans, not of wits and scholars. Merchants and artisans are men who handle *things* and who presumably handle words with a similar probity. Wits and scholars handle nothing more substantial than ideas. Journalism seemed guaranteed by the plain style. Handbooks and copy editors now teach journalists how to write plainly, that is, in such a manner that they will be trusted. You get yourself trusted by artifice.

Plain style is a populist style and one that suited writers like Swift, Mencken, and Orwell. Homely diction is its hallmark, also one-two-three syntax, the show of candor and the artifice of seeming to be grounded outside language in what is called fact—the domain where a condemned man can be observed as he silently avoids a puddle and your prose will report the observation and no one will doubt it. Such prose simulates the words anyone who was there and awake might later have spoken spontaneously. On a written page, as we've seen, the spontaneous can only be a contrivance.

So a great deal of artifice is being piled on, beginning with the candid no-nonsense observer. What if there was a short circuit: no observation, simply the prose? Whenever that is suggested, straightforward folk get upset. But they were never meant to think about it, any more than airplane passengers are meant to brood about what holds them aloft—thin air. The plain style feigns a candid observer. Such is its great advantage for persuading. From behind its mask of calm candor, the writer with political intentions can appeal, in seeming disinterest, to people whose pride is their no-nonsense connoisseurship of fact. And such is the trickiness of language that he may find he must deceive them to enlighten them. Whether Orwell ever witnessed a hanging or not, we're in no doubt what he means us to think of the custom.

His masterly plain style came to full development in *Homage to Catalonia* (1938), an effort to supply a true account of the Spanish Civil War, about which the communists, his one-time allies, were fabricating a boilerplate account. Though their ostensible enemies were the so-called fascists, the communists also figured that much trouble was being made by treasonable Trotskyists, whom they accused of making an alliance with the fascists to subvert genuine revolution. It was in communist so-called news of the mid-1930s that Orwell first discerned newspeak, which penetrated not only *The Daily Worker* but respectable London papers like *The News Chronicle*. What they printed was the news, and it was believed. How to counter what was believed?

Why, by the device of the firsthand observer, a device as old as Defoe, who used it in *A Journal of the Plague Year* to simulate persuasive accounts of things he could not have seen. When newspeak indulges in sentences like "Barcelona, the first city in Spain, was plunged into bloodshed by *agents provocateurs* using this subversive organization," your way to credibility is via sentences like this: "Sometimes I was merely bored with the whole affair, paid no attention to the hellish noise, and spent hours reading a succession of Penguin Library books which, luckily, I had bought a few days earlier; sometimes I was very conscious of the armed men watching me fifty yards away." After you've established your credentials like that, your next paragraphs can ignore the newspeak utterance as mere academic mischief. And it literally doesn't matter whether you read Penguins in Spain or not.

Orwell was alert to all of English literature, from Chaucer to *Ulysses*. A source for the famous trope about some being more equal than others has been found in *Paradise Lost*. He had studied Latin and Greek, and once, when hard up, he advertised his readiness to translate from anything French so long as it was from after 1400 A.D. Yet he is identified with an English prose that sounds native, a codifying of what you'd learn by ear in the Wigan of Orwell's *Road to Wigan Pier,* or in any other English working-class borough. Newspeak, as he defined it in *Nineteen Eighty-Four,* seems to reverse the honesty of all that: War is peace, freedom is slavery, $2 + 2 = 5$. Political discourse being feverish with newspeak, he concocted his plain style to reduce its temperature.

We are dealing now with no language humans speak, rather with an implied ideal language the credentials of which are moral, a language that cleaves to things and has univocal names for them. "Cat" is cat, "dog" is dog. That, in Swift's time, had been a philosopher's vision, and in *Gulliver's Travels* Swift had derided philosophers who, since words were but tokens for things, saved breath and ear by reducing their discourse to a holding up of things.

Examine Orwell's famous examples, and discover an absence of opposite *things*. "War" is not war the way "cat" is cat, nor is "freedom" freedom the way "dog" is dog. Such abstractions are defined by consensus. Once we've left behind "cat" and "dog" and "house" and "tree," there are seldom things to which words can correspond, but you can obtain considerable advantage by acting as if there were. The plain style, by which you gain that advantage, seems to be announcing at every phrase its subjection to the check of experienced and namable things. Orwell, so the prose says, had shot an elephant. Orwell had witnessed a hanging. Orwell at school had been beaten with a riding crop for wetting his bed. The prose says these things so plainly that we believe whatever else it says. And none of these things seem to have been true.

We should next observe that Orwell's two climactic works are frank fictions—*Animal Farm* and *Nineteen Eighty-Four*. In a fiction you address yourself to the wholly unreal as if there were no doubt about it. In *Animal Farm* we're apprised of a convention at which pigs talk to one another. Except for the fact that we don't credit pigs with speech, we might be attending to a report of a county council meeting. (And observe which way the allegory runs; we're not being told councilors are pigs.)

It is clarifying to reflect that the language of fiction cannot be told from that of fact. Their grammar, syntax, and semantics are identical. So Orwell passed readily to and fro between his two modes, reportage and fiction, which both employ the plain style. The difference is that the fictionality of fiction offers itself for detection. If the fiction speaks political truths, it does so by allegory.

That is tricky, because it transfers responsibility for what is being said from the writer to the reader. Orwell's wartime BBC acquaintance William Empson warned him in 1945 that *Animal Farm* was liable to misinterpretation, and years later Empson himself provided an object lesson when he denied that *Nineteen Eighty-Four* was "about" some future communism. It was "about," Empson insisted, as though the fact should have been obvious, that pit of infamy, the Roman Catholic Church. One thing that would have driven Empson to such a length was his need to leave the left unbesmirched by Orwell and Orwell untainted by any imputation that he'd besmirched the left. Empson summoned Orwell's shade to abet the hysteria he was indulging at the moment. He was writing about *Paradise Lost,* contemplation of which appears to have unsettled his mind.

Now, this is an odd place for the plain style to have taken us, a place where there can be radical disagreement about what is being said. "A close, naked, natural way of speaking," Sprat had written; "positive expressions, clear senses . . . thus bringing all things as near to the mathematical plainness as they can." That is terminology to depict a restored Eden, before both Babel and Cicero, when Adam's primal language could not be misunderstood, when words could not possibly say (as Swift mischievously put it) "the thing that was not." That was when Adam delved and Eve span, and they both had the virtues of merchants and artisans—as it were, Wigan virtues.

But the serpent misled them, no doubt employing the high style, and what their descendants have been discovering is that not even the plain style can effect a return to any simulacrum of Paradise. Any spokesman for political decencies desires the Peaceable Kingdom. Books like *Animal Farm* and *Nineteen Eighty-Four* can go awry. *Gulliver's Travels* does; it ends with the hero-narrator longing vainly to be a horse. What the masters of the plain style demonstrate is how futile is anyone's hope of subduing humanity to an austere ideal. Straightness will prove crooked, gain will be short-term, vision will be fabrication and simplicity an intricate contrivance. Likewise, no probity, no sincerity, can ever subdue the inner contradictions of speaking plainly. These inhere in the warp of reality, ineluctable as the fact that the square root of two is irrational. Swift is called mad, Orwell was reviled for betraying the left, and by divulging the secret of the root of two, which was sacred to the Pythagoreans, a Greek named Hippasos earned a watery grave.

8

The New Journalism and the Image-World

David Eason

By the end of the 1960s, the doctrine of representation had crumpled in linguistics, philosophy, and even literary criticism.[1] Cultural criticism focused on how the self might find its bearing in a society characterized by a breakdown in consensus about manners and morals and by the permeation of everyday life by a mass-produced image-world.[2] The relationship between image and reality was not holding, and the impact of this development was reflected in journalism as well. A New Journalism emerged in magazines and in books to give shape to many of the cultural changes while revitalizing reporting as a form of storytelling.

The New Journalism took its energy from the recognition of society as a tableau of *interesting* races, age groups, subcultures, and social classes and the *detachment* of the self from various conventional sources of identification. New Journalists found many of their stories in trends and events already transformed into spectacles by the mass media. They used the image-world as a background for investigations into the significance of the counterculture, political campaigns and conventions, prison rebellions, murders, executions, and other spectacular events.

The reports most often characterized as New Journalism share some important characteristics.[3] They usually focus on events as symbolic of a cultural ideology or mythology, emphasize the world view of the individual or group under study, and show an absorption in the aesthetics of the reporting process in texts that read like novels or short stories. Despite these similarities, the reports can be best understood as embodying two different ways of responding to the problem of social and cultural diversity and of locating the reporter in regard to the traditions of journalism and the broader history of American society. I will term these approaches *realist* and *modernist*.[4] Realist texts, seen most clearly in the reporting of Tom Wolfe, Gay Talese, and Truman Capote, organize the topic of the report as an object of display, and the reporter and reader, whose values are assumed and not explored, are joined in an act of observing that assures conventional ways of understanding still apply. In contrast, modernist texts, most clearly reflected in the reporting of Joan Didion, Norman Mailer, and Hunter Thompson, describe what it feels like to live in a world where there is no consensus about a frame of reference to explain "what it all means." The reports focus instead on the contradictions that emerge at the intersection of various maps of experience. The phenomenological impetus of the modernist reports is revealed in the multilayered questioning of communication, including that between writer and reader, as a way of making a common world, and by the hesitancy to foreclose the question, "Is this real?" by invoking conventional ways of understanding.

The differences between the two forms of reporting are visible in three dimensions: (1) Realist reports organize experience in terms of the traditional duality between image and reality. The reporter must go beyond the image in order to reveal the reality it hides. Reality, though elusive, nonetheless waits to be discovered. Modernist reports describe a world where image and reality are intertwined to such an extent that they call commonsensical views into question. (2) Realist reports suggest the priority of cultural categories such as observed and lived experience. The reports describe observing as a professional act that poses only manageable ethical problems. In modernist reports, such assumptions are examined as ways of legitimizing ethical decisions. (3) Realist reports reflect faith in the capability of traditional models of interpretation and expression, particularly the story form, to reveal the real. Although the reports acknowledge cultural relativism in their attention to the various symbolic worlds of their subjects, this awareness is not extended to the process of reporting, which is treated as a natural process. Modernist reports call attention to reporting as a way of joining writer and reader together in the creation of reality. Narrative techniques call attention to storytelling as a cultural practice for making a common world.

IMAGES AND REALITIES

The realist impulse, as Tom Wolfe describes it, is "to show the reader *real life*—'Come here! Look! This is the way people live these days! These are the things they do!' "[5] In *The Electric Kool-Aid Acid Test,* Wolfe described the process of interpretation whereby symbols and experience are connected. When he meets Ken Kesey and his Merry Pranksters, Wolfe puzzles over their style of life.

> Figuring out parables, I look around at the faces and they are all watching Kesey and, I have not the slightest doubt, thinking: and *that's what the cops-and-robbers game does to you.* Despite the skepticism I brought here, *I* am suddenly experiencing *their* feeling. I am sure of it. I feel like I am in on something the outside world, the world I came from, could not possibly comprehend.[6]

Making the scene comprehensible also involved relating it to a social, cultural, or historical framework. Gay Talese explains the interaction between his subjects and a changing social context by placing them in a historical framework.[7] Truman Capote infuses his account of the murder of a Kansas farm family with psychological and cultural explanations.[8] Wolfe, the most ambitious of the realists, furnishes the most elaborate explanations, explanations that link the contemporary to a well-ordered, nonthreatening past that promises to extend into the future. Underlying the ambiguous surfaces of contemporary reality, Wolfe's reports suggest, is a cultural pattern working itself out. The drug culture, symbolized by Ken Kesey and his Pranksters, may appear to be a multitude of styles and symbols with no apparent meaning, but it is actually only a new manifestation of an ancient religious impulse and the group an elementary form of religious life. The "radical chic" of

New York socialites may appear to be a contradictory social rite for the elite to finance its own destruction, but it is actually only a particularly humorous manifestation of upper-class consumption traceable to the nineteenth century. The diversity of California lifestyles may suggest the loss of a common culture, but they are actually only diverse expressions of a culture transformed by economic expansion that spread the idiosyncratic lifestyles of the upper-class throughout society.[9]

While realism takes its energy from an image-world that obscures the subjective realities of diverse subcultures, exploring these alternative realities poses no threat to the reporters' faith that they can discover, comprehend, and communicate the real. Likewise these discrepant realities pose no threat to the society's identity. Realist texts are organized around naturalizing discrepant views of reality within their own narrative conventions. Social reality may indeed be bizarre, but it poses no threat to established ways of knowing and communicating. The disorder of society is only apparent and the reporter is still able to state, "That's the way it is."

While realism assures its readers that traditional ways of making sense still apply in society, modernist texts describe the inability of traditional cultural distinctions to order experience. Didion introduces her first collection of reportage as a confrontation with the "evidence of atomization, the proof that things fall apart." She began working on the title piece of *Slouching Towards Bethlehem* after a time when she "had been paralyzed by the conviction that writing was an irrelevant act, that the world as I had understood it no longer existed." The report, as well as many of the others in the volume, is an attempt "to come to terms with disorder."[10]

Modernist texts reveal that coming to terms with disorder is not synonymous with making events comprehensible for the outside world. They describe the image-world as a realm that blurs traditional distinctions between fantasy and reality. Didion described a land where people have drifted away from the security of tradition in order "to find a new life style . . . in the only places they know to look: the movies and the newspapers."[11] Mailer sees the counterculturists of the 1960s as "assembled from all the intersections between history and the comic books, between legend and television, the Biblical archetypes and the movies."[12] Michael Herr describes the Vietnam War as one where a commander would risk lives solely for "a little ink."[13] Didion suggests that the true significance of the California murder trial of Lucille Miller was that it made public the ways in which image and reality had become fatalistically intertwined in private lives:

> What was most startling about the case that the State of California was preparing against Lucille Miller was something that had nothing to do with law at all, something that never appeared in the eight-column afternoon headlines but was always there between them: the revelation that the dream was teaching the dreamers how to live.[14]

The dreamlike nature of American reality in the 1960s and early 1970s is the major theme of Hunter Thompson's reportage. Whether focusing on leisure, sports, or politics, Thompson describes a culture where the real has become so saturated with the fantastic that knowledge and ethics have become problematic in new ways. *Fear and Loathing in Las Vegas* describes Las Vegas as the metaphorical center of the image-world, a place where "everydayness" is created from surreal transformations of movies and television shows. Against this backdrop, Raoul Duke and Dr. Gonzo search for "the main story of our generation," the transformation of the American Dream into a neon nightmare organized around the hedonistic pursuit of "the now."[15]

In *The Image* Daniel Boorstin suggests nostalgically that Americans should return to previous distinctions between image and reality, recognizing that extravagant expectations make reality more remote. The modernist texts of New Journalism argue that such a return is impossible. The disorder of contemporary society is rooted in the interaction of images of reality and the reality of images that creates a realm where there is no exit.

OBSERVING AND LIVING

Human interest reporting traditionally requires a form of psychic distance that allows the reporter to find the "facts" in the "inner experiences" of the subject.[16] Ortega y Gasset gives an example of this kind of distance: A famous man is dying. Around his bed are a number of people including his wife, a doctor, and a reporter, each of whom participates in the scene differently. The wife is totally involved in the situation. She does not behold the scene, Ortega y Gasset argues, she lives it. The doctor is involved in the scene, but only partially, responding to the man's needs through his professional self. Likewise, the reporter participates in the scene through his professional self. Whereas the doctor's professional code demands he be involved, the reporter's demands that he be aloof. "He does not live the scene, he observes it."[17] Still, because within a matter of hours he must construct a compelling narrative that will move his readers by placing them in the scene, he observes in a particular way.

Reporters must be simultaneously near and far from their subjects. They must vicariously penetrate experiences while holding an aesthetic distance that allows them to see the story in those experiences. This distinction between lived and observed experience is a fundamental distinction for human interest reporting, which sometimes requires reporters to become involved in the lives of their subjects. Since a reporter who treats each assignment as an occasion for self-analysis is likely to have difficulty legitimizing being a reporter over time, the distinction between lived and observed experience is a routine assumption that makes possible the daily telling of stories about the lives of other people.

Realist reports maintain the interpretive stance of human interest reporting. The distinction between lived and observed experience is crucial to the form's principle of realism that real life is someone else's. "The subjectivity that I value in the good examples of New Journalism," Wolfe says in an interview, "is the use of techniques to enable the writer to get the subjective reality—not his own—but the character he's writing about."[18] Talese makes a similar point, using the metaphor of the "movie director" to describe the process. "I'm like a director, and I shift my own particular focus, my own cameras, from one to the other. . . . I find that I can then get into the people that I am writing about and I just shift." Paradoxically, the distance of the director allows a more intimate portrait. In preparing a profile on Frank Sinatra, Talese did not interview him. Defending his style of research, Talese says, "I don't think I could have asked him—nor do I think he could have answered—those questions in any way that would have been as revealing of himself as I was able to gather by staying a bit of a distance away, *observing* him and overhearing him and watching those around him react to him."[19]

In order to keep the subjects within the domain of the story type, the realist must maintain the world of the other person as a discrete province of meaning, similar to that experienced in a film or play. Wolfe, for instance, cautions about the hazards of confronting the subject as more than a story type:

> If a reporter stays with a person or group long enough, they . . . will develop a personal relationship of some sort. . . For many reporters this presents a more formidable problem than penetrating the particular scene in the first place. They become stricken with a sense of guilt, responsibility, obligation. . . . People who become overly sensitive on this score should never take up the new style of journalism.[20]

In this argument, and others, Wolfe connects the practices of realism with the traditions of human interest reporting. Thompson is sensitive to this connection when he writes of Wolfe, "The only thing new and unusual about Wolfe's journalism is that he's an abnormally *good* reporter . . . *good* in the classical—rather than the contemporary—sense."[21]

The contemporary sense to which Thompson alludes involves calling into question many of the traditional assumptions of the reporting process. Whereas realist reports, like other forms of journalism, reveal the act of observing to be a means to get the story, modernist reports reveal observing as an object of analysis as well. The well-ordered social dramas described in realist reports become disrupted spectacles in modernist texts where the roles of actor and spectator are no longer clearly defined. Modernist reports describe observing as a social form for organizing experience and explore how the act of observing supports what Wicker calls "the ethic of the press box."[22] Such an ethic suggests that to become an observer is to see social reality as composed of active participants, who must take responsibility for their acts, and passive spectators, who bear no responsibility for what they watch.

Although Mailer, Didion, and Thompson focus on the relationship of observed and lived experience, the most extended treatment of the relationship is John Gregory Dunne's *Vegas* (1974). In *Vegas,* Dunne explores the motivation for reporting and considers the ethical dilemmas implicit in creating worlds of experience for anonymous readers. Dunne goes to Las Vegas because he is depressed and believes that reporting will be therapeutic. He plans to write a realist account of the underside of Las Vegas and he discovers three characters for his book: a young woman who works as a prostitute at night and attends beauty school during the day, a retired police officer from the Midwest who runs a private detective agency, and a small-time lounge comic who dreams of being a star. Although he desires to see the three as windows to the underlying cultural reality of Las Vegas, the characters function instead as mirrors for self-examination.

Dunne concludes that he is a voyeur who lives his life through others in order to avoid dealing with his own problems. He previously legitimized being an observer in life on moral grounds.

> I was good at it, and imagined I had an empathy for those I observed. I liked to think, I told myself then, and was still telling myself that summer in Vegas, that I could learn something about myself from the people whose lives I intruded upon, indeed that this was the reason I had taken up residence in an apartment behind the Strip.[23]

In Las Vegas Dunne comes to believe that reporting is less a search for stories containing morals about how to live than a search for an objectification of his own fantasy life. For example, his fascination with the young prostitute is less a quest for an understanding of her world than a socially legitimate way of having a prostitute and "losing" the self:

> Reporting anesthetizes one's own problems. There is always someone in deeper emotional drift, or even grift, than you, someone to whom you can ladle out understanding as if it were a charitable contribution, one free meal from the psychic soup kitchen, just that one, no more, any more would entail responsibility, and responsibility is what one is trying to avoid in the first place.[24]

Dunne describes the reporter as one who quests for a variety of experiences, believing that all knowledge can make one more morally aware. Michael Herr makes this point explicit when he writes of the Vietnam War, "I went there behind the crude but serious belief that you had to be able to look at anything . . . it took the war to teach it, that you were as responsible for everything you saw as you were for everything you did."[25] Reporting, in the modernist accounts, is less a realm of certified activities than an activity in need of new intellectual and moral justifications.

STORIES AND EXPERIENCES

Both forms of New Journalism reflect an absorption in aesthetic concerns. The emphases of the two forms, however, are distinctive. Realist reports represent style as a communication technique whose function is to reveal a story that exists "out there" in real life. Modernist texts represent style as a strategy for conceiving as well as revealing reality. In realist reports, the dominant function of the narrative is to reveal an interpretation; in modernist reports it is to show how an interpretation is constructed.

The realist confronts narrative construction as a problem of mediating between the experience of the subject and the reader. All relationships necessary to complete the reporting process are subsumed in the creation of a symbolically unified set of scenes. What gives the report its novelistic quality is the invisible camera eye of the narrator that can record all of the objective details of the scene, then move in and out of the characters' experiences. The reporter's own experiences, the inevitable "jump cuts" of the reporting process, are either deleted or used as "cutaways" to unify the narrative. Wolfe justifies dealing with communication as a technical matter through what he terms "the physiology of realism." Print is an indirect medium that does not image what it represents. It operates by association, calling forth from the reader's experience typifications that provide a context of meaning for the story. "The most gifted writers are those who manipulate the memory sets of the reader in such a rich fashion that they create within the mind of the reader an entire world that resonates with the reader's own real emotions."[26]

Although stories are told and not lived, "thinking as usual" transforms a way of communicating into a taken-for-granted reality.[27] The narrative line is believed to exist in the events, not in the story. When experience fails to fit the pattern and makes the relativity of the narrative line conscious, a gap occurs. Didion makes explicit the role of this experiential gap in the modernist texts when she writes in *The White Album:*

I was meant to know the plot, but all I knew was what I saw: flash pictures in variable sequence, images with no "meaning" beyond their temporary arrangement, not a movie but a cutting-room experience. In what would probably be the middle of my life I wanted still to believe in the narrative and in the narrative's intelligibility, but to know that one could change the sense with every cut was to begin to perceive the experience as rather more electrical than ethical.[28]

In modernist texts, the story that is told is not one discovered out there in the world, but is instead the story of the writer's efforts to impose order on those events.

The most complex treatment of the relationship between the forms of narrative and the order of experience is Mailer's *The Armies of the Night*. The report focuses on an antiwar march on the Pentagon in 1967 and was created against a background of mass media reports that established a conceptual framework for understanding the event.[29] The book removes the march from this taken-for-granted context and describes it as a mysterious, ambiguous event whose meaning is not readily apparent. Mailer argues that the event as reported in the mass media is paradigmatic of the routines and assumptions of the news media and instead focuses on the event in a larger cultural context as "a paradigm of the twentieth century" that occurs "in the crazy house of history." The absurdity of the event is bound up in the conditions for political action in a technological society that drives a wedge between the "symbolic" and "real" dimensions of events. Although the march is a real event that results from the commitment of individuals to stop the war in Vietnam, it is organized in terms of the way it will be seen in society. The expressive dimension of event, reflected in the willingness of individuals to face personal injury and jail, is overwhelmed by its communicative dimension. "A protest movement which does not grow loses power every day, since protest movements depend upon the interest they arouse in mass media. But the mass media are interested only in processes which are expanding dramatically or collapsing." The need to appear to be a growing force demands a "revolutionary aesthetic," a consumption of the techniques of media production that threatens to transform protesters into actors.[30]

Mailer is in the front ranks of the protesters because he is a celebrity, guaranteed to bring media coverage. His relationship to his own image exemplifies the transforming powers of mass media and the disjunction between the symbolic and the real that pervades modern culture. "Mailer's habit of living—no matter how unsuccessfully—with his image, was so engrained by now, that like a dutiful spouse he was forever consulting his better half."[31]

The structure of *The Armies of the Night* makes the realm of appearances created by the mass media problematic. Mailer substitutes a narrative strategy designed to reveal the disjunction between reported and experienced reality for the predictable pattern of news reports. The report is self-consciously punctuated with comments about narrative construction that reveal the transforming powers of all narratives. Essential to this strategy is the two-part organization of the book, "History as a Novel; the Novel as History," which emphasizes not events but the interaction of narrative forms and events. History, Mailer argues, deals with the realm of behavior. As a symbolic structure it places the communicative function in the foreground. The novel deals with the realm of the "emotional, spiritual, psychical, moral, existential, or supernatural," placing the expressive dimension of the symbolic act in the foreground.[32] The two-part strategy of *The Armies of the Night* emphasizes that both social life and the report are constructions. The historian attempting to penetrate the realm of appearances surrounding the march can only weigh the various official versions that the event produced. These public versions, however, do not mirror the private realities of the march. In the face of such contradictions, the historian must become a novelist who invents the meaning of the event. Unlike the realist who penetrates the image to find the reality of the other, Mailer argues that the image is impenetrable, that significance is privatized in the "novelistic" lives that make up the map of contemporary reality. Confronting such a dilemma, the individual cannot find the facts but must invent an interpretation in an act of self-definition.

The mythology of a technological society, Mailer argues, is that disembodied facts communicate. *The Armies of the Night* takes the demystification of this mythology as its object. It seeks to create for the reader the disorder that interpretations transform, the experiential contradictions that usually remain outside the text. The book is an account of experience that is, in Didion's words, more electrical than ethical but from which an ethical stance must be created.

In the realist text, the diversity of contemporary society is interesting but not threatening. Underlying the disparate images of culture are patterns that appear again and again in history. People may wear a variety of masks, but the spectators, the reporter and the reader, can still distinguish the artificial from the real. The relativity of human actions that motivates the reports poses no threats to traditional ways of comprehending and expressing reality. In the modernist text, contemporary society is all flux. The reporter has no privileged position, in Mailer's words, no tower to look down upon human actions. The changes that the reports chronicle are not merely changes occurring out there in the world and in the consciousness of the other, they are pervasive in society. Modernist reports transform what is taken for granted in writing and reading a report into an object of analysis. Reporter and reader are implicated in the social changes themselves. Meaning, these texts argue, is not something that exists out there independent of human consciousness, but something that is created and recreated in acts of interpretation and expression.

ON DOCUMENTING EXPERIENCE

Writing in *Mythologies* in the 1950s, Roland Barthes saw the problem that dominates the modernist reports in a somewhat different light:

> The fact that we cannot manage to achieve more than an unstable grasp of reality doubtless gives the measure of our present alienation: we constantly drift between the object and its demystification, powerless to render its wholeness. For if we penetrate the object, we liberate it but we destroy it; and if we acknowledge its full weight, we respect it, but we restore it to a state which is still mystified. It would seem that we are condemned for some time yet always to speak *excessively* about reality.[33]

The modernist form of New Journalism is a mode of excessive speech that finds its home in the space between realism and relativism. It is a form of expression for a culture organized around the proliferation of realistic reports that simultaneously experiences the loss of standards for realism. That such a feeling is not pervasive is indicated by the realist form. The modernist texts, however, attest to just how fragile notions of realism are in a self-conscious culture, and to the impossibility, in some historical moments, of speaking excessively about reality.

In giving shape to the social and cultural changes of the late 1960s and early 1970s, the New Journalism extended the principles of realism that have guided journalistic writing since the nineteenth century. In its most original forms, the New Journalism also imported into the realm of reporting the concerns that have occupied the leading modernist writers. If these diverse realist and modernist texts share a common identity as New Journalism, it is in the spirit of experimentalism that, on the one hand, pushed the conventions of realism to order emerging forms of subjectivity, and, on the other, dramatized the gaps between those very conventions and forms of subjectivity they seemed incapable of containing. As almost a century of modernist experimentation shows, however, experience does settle into new containers and these containers, in turn, can come to seem as incapable of grasping reality as those they displaced. We long for the new, that story that seems just beyond us, but we also long for the comfort of the predictable, the story that pulls us along in familiar ways. We find our world, and ourselves, in these mirrors that mix us up and distort us and then give us back to ourselves in moments of clarity.

NOTES

1. This essay originally appeared in *Critical Studies in Mass Communication* I(1):51–65. In preparing the manuscript for this collection, I have made some revisions: shortening the manuscript, simplifying some terms, and extending the conclusion by a paragraph.

2. See, for instance, Daniel Boorstin, *The Image* (New York: Harper & Row, 1964); R. D. Laing, *The Politics of Experience* (New York: Ballantine Books, 1967): Marshall McLuhan, *Understanding Media* (New York: McGraw-Hill, 1964); and Guy Debord, *Society of the Spectacle* (Detroit: Black and Red, 1970).

3. On the New Journalism see Everette Dennis and William L. Rivers, *Other Voices: The New Journalism in America* (San Francisco: Canfield Press, 1974); Michael Johnson, *The New Journalism* (Lawrence: University of Kansas Press, 1971); John Hollowell, *Fact and Fiction: The New Journalism and the Nonfiction Novel* (Chapel Hill: The University of North Carolina Press, 1977); James E. Murphy, "The New Journalism: A Critical Perspective," *Journalism Monographs* 34

(May 1974); Ronald Weber, *The Literature of Fact* (Athens: Ohio University Press, 1980); John Hersey, "The Legend on the License," *The Yale Review* 70:1–25 (Autumn 1980); John Hellman, *Fables of Fact: The New Journalism as New Fiction* (Urbana: University of Illinois Press, 1981); Joseph M. Webb, "Historical Perspective on New Journalism," *Journalism History* 1: 38–60 (Summer 1974); Mas'ud Zavarzadeh, *The Mythopoeic Reality: The Postwar American Nonfiction Novel* (Urbana: University of Illinois Press, 1976); and my "New Journalism, Metaphor and Culture," *Journal of Popular Culture* 15(4): 142–49 (Spring 1982). Collections reflecting the varied critical response are "The New Journalism," *Journal of Popular Culture* 9: 99–249 (Summer 1975); and Ronald Weber, ed., *The Reporter as Artist: A Look at the New Journalism Controversy* (New York: Hastings House, 1974).

4. In the earlier version of this essay, I termed these approaches "ethnographic realism" and "cultural phenomenology." Here, I am following the simpler terminology of George E. Marcus and Michael J. Fischer in *Anthropology as Cultural Critique* (Chicago: University of Chicago Press, 1986), pp. 67–73.

5. *The New Journalism* (New York: Harper & Row, 1973), p. 33.

6. (New York: Bantam Books, 1968), p. 25.

7. See *The Kingdom and The Power* (New York: World Publishing Co., 1966); *Honor Thy Father* (New York: Fawcett-World, 1972); and *Thy Neighbor's Wife* (New York: Doubleday, 1980).

8. *In Cold Blood* (New York: Random House, 1965).

9. *The Electric Kool-Aid Acid Test; Radical Chic and Mau-Mauing the Flak Catchers* (New York: Farrar, Straus & Giroux, 1970); *The Pump House Gang* (New York: Farrar, Straus & Giroux, 1968); and *The Kandy-Kolored Tangerine-Flake Streamline Baby* (New York: Farrar, Straus & Giroux, 1965). See also *The Right Stuff* (New York: Farrar, Straus & Giroux, 1979), which sharply distinguishes between the image and reality of the astronauts.

10. (New York: Delta, 1968), pp. xi–xii.

11. ibid., p. 4.

12. *The Armies of the Night: History as a Novel; The Novel as History* (New York: New American Library, 1968), p. 107.

13. *Dispatches* (New York: Alfred A. Knopf, 1977), p. 7.

14. *Slouching Towards Bethlehem*, p. 17.

15. Hunter S. Thompson, *Fear and Loathing in Las Vegas: A Savage Journey to the Heart of the American Dream* (New York: Popular Library, 1971). See also *Fear and Loathing on the Campaign Trail* (San Francisco: Straight Arrow Books, 1973) and *The Great Shark Hunt* (New York: Fawcett, 1979).

16. Helen M. Hughes, *News and the Human Interest Story* (Chicago: University of Chicago Press, 1940), p. 101.

17. Ortega y Gasset, J. *The Dehumanization of Art.* (Princeton: Princeton University Press, 1948), pp. 14–17.

18. Joe David Bellamy, *The New Fiction: Interviews with Innovative American Writers* (Urbana: University of Illinois Press, 1974), p. 85.

19. John Brady, "Gay Talese: An Interview," in Weber, *The Reporter as Artist*, pp. 94–97.

20. *The New Journalism*, p. 51.

21. *The Great Shark Hunt,* p. 108.

22. Tom Wicker, *A Time to Die* (New York: Quadrangle, 1975), p. 37. Also see Susan Sontag, *On Photography* (New York: Delta, 1978) and Douglas Birkhead, "An Ethics of Vision for Journalism," *Critical Studies in Mass Communication* 6:283–94 (December 1989).

23. *Vegas: A Memoir of a Dark Season* (New York: Random House, 1974), p. 208.

24. ibid., p. 14.

25. *Dispatches,* p. 20.

26. *The New Journalism,* p. 48.

27. See Alfred Schutz, "The Stranger," *Collected Papers, Vol. II: Studies in Social Theory* (The Hague: Martinus Nijhoff, 1964), pp. 91–96.

28. *The White Album* (New York: Simon and Schuster, 1979), p. 13.

29. See also *Miami and the Siege of Chicago* (New York: New American Library, 1968) and *Of a Fire on the Moon* (Boston: Little, Brown & Co., 1970). Although *The Executioner's Song* (Boston: Little, Brown & Co., 1979) also focuses on a mass-mediated event, Mailer's approach raises ethical dilemmas common to the realist text.

30. *The Armies of the Night,* p. 257.

31. ibid., p. 182.

32. ibid., p. 281.

33. Roland Barthes, *Mythologies* (New York: Hill and Wang, 1972), Annette Lavers, trans., p. 159.

9

John McPhee Balances the Act

Kathy Smith

> "You know," he explained to Lapham, "that we have to look
> at all these facts as material, and we get the habit of classify-
> ing them. Sometimes a leading question will draw out a whole
> line of facts that a man himself would never think of."[1]
>
> William Dean Howells, *The Rise of Silas Lapham*

John McPhee has earned a reputation for poised writing, in two
senses of the word. His polished and confident style balances deftly
on the verge. The respect bestowed on him by colleagues and
journalism professors nationwide can be attributed to this poise,
which has been described for the most part in a New Critical vocabulary more generally applied to fiction: The McPhee style consists of well-crafted sentences, fresh and spirited metaphors, consis-
tency of mood and tight organizational control, strong narrative
voice and an uncanny form/content justification, an organic unity
that reaffirms the aesthetic value of parts fitting the whole.[2] In
1965, *New Yorker* editor William Shawn hired McPhee as a staff
writer following publication of his profile on basketball player Bill
Bradley. This story, "A Sense of Where You Are," will serve as the
focusing text for my comments on McPhee's established place as a
writing subject and his work's place in what might be called the
new contract of literary journalism.

While my use of the two concepts, *writing subject* and *contract,*
will take on context in the course of the chapter, a brief explanation may help guide the following discussion. The choice of the
phrase *writing subject* is a deliberate one meant to heighten the

awareness that a writer or author at once subjects, is a subject of, and is subject to the material composed in the act of writing itself. The conventions of journalistic writing impose boundaries meant to contain and maintain objectivity and to limit rhetorical and narrative choices. When one calls oneself a journalist, therefore, one takes up a judicial position in regard to differentiating between fact and fiction. However, as a writer in the more general sense, and as a manipulator of the material he fashions into story, McPhee constantly crosses and tests those boundaries. McPhee poses as a paradoxical figure, both crafty and innocent, who reveals himself in the position he takes up within the contract. As he does so he questions the meaning of the two categories of fact and fiction, a claim I hope to substantiate later on.

When we regard the author as a "subject" of writing itself, the effect on the contract (understood in journalism as a tacit agreement with the reader and the ostensible story subject to make a true accounting) is to substitute anxiety for assurance. If we expect truth and verifiability from this true story, then we must impose artificial restraints on the story line. This is where journalism's codes come in, protecting the true from the story. McPhee may well be regarded by most nonfiction writers as a master tactician when it comes to maintaining both the integrity of the journalism profession and the art of writing. When the two collide, the reader–writer contract can be reassessed to discover what has been understated. If the narrative is labeled *nonfiction,* then the obvious assumption is that the event takes a privileged place in its relationship to writing. The new contract written by literary journalists, among whom McPhee must count, disturbs this assumption by calling into question the priority of the event over writing. One might indeed say that in the new contract writing always precedes the event.

In his introduction to *The John McPhee Reader,* William Howarth calls attention to McPhee's "transparency" in social settings and its importance to his role as reporter:

> A good reporter, he moves through crowds easily, absorbing names, details, snatches of talk. He inspires confidence, since people rarely find someone who listens that carefully to them. Around Princeton, old neighbors and schoolmates remember him fondly, if not well. He cultivates a certain transparency in social relations, a habit de-

rived from practicing his craft. To see and hear clearly, he keeps his eyes open and mouth shut.[3]

Howarth continues—for the bulk of the introduction is concerned with McPhee's process of composition and his theory of writing— by noting McPhee's almost fanatical preoccupation with controlling the story's structure. Howarth's description points to an interesting conflict, one that seems endemic to the craft of literary journalism: the reporter adopts an attitude of cozy selflessness, a kind of partisan nonpartisanship; he takes on the solid and respectable aspect of a nonjudgmental witness in order to be privy to the "essential" subject. Later, he metamorphoses into a willful usurper or supersleuth, imposing order, center, logic, and meaning.

The story, one is led to imagine, both writes itself—possesses a natural internal order—and, at the same time, needs to be shaped. McPhee's titles are revealing: *A ROOMFUL of Hovings, LEVELS of the Game, ENCOUNTERS with the Archdruid, PIECES of the Frame.* The metaphors are spacial and structural. They produce an arrangement that seems to partake of the subject's own particular structure. They illustrate McPhee's desire to control and make sense of the real in a conventionally novelistic way, where plot, character, setting, and mood develop in a framework based not on a traditionally historical model of temporal progression, like a chronicle or a list, but on a narrative one, where the author takes liberties with the order and structure of events so that the story advances strategically along thematic lines. On the written level, the synthesis of the world and the written word results in something like what Howarth calls a *true replica* in his introduction. Of course the phrase has much more resonance than Howarth can begin to discuss there.

A provocative indirection may discover for us a useful analogue to this notion of the true replica. In *Image, Music, Text,* French literary critic Roland Barthes writes:

> The function of narrative is not to "represent," it is to constitute a spectacle, . . . Narrative does not show, does not imitate . . . "What takes place" in a narrative is from the referential (reality) point of view literally *nothing;* "What happens" is language alone, the adventure of language, the unceasing celebration of its coming.[4]

Seen from Barthes's point of view, the news story as narrative cannot be fully apprehended merely as an *effect* of an event that, having taken place in time, is transposed into a new time of linguistic adventure. The replica certainly bears resemblance to a truth that the author monitors, anatomizes, and then reproduces making careful rhetorical choices. These choices, however, are not based on a model of reality established through recitation or observation of fact, or not merely.

The assumption one makes with narrative in general is that the more appropriate the literary figure chosen to represent a real phenomenon, the more true the replica. To avoid the trap of this circular logic, Barthes proposes the "spectacle" of language itself, which serves as its own referent, its own remark on itself. It may seem almost perverse to read McPhee, of all people, through Barthes. But Barthes's articulation of the writing adventure is clearly relevant to any discussion of labels—both McPhee's own and those used by others to describe his work. Barthes's theory of language in its limited use here will provide a means of seeing something that McPhee cannot readily reveal about how he prepares us for reading.

Labels and titles obviously cause us to think in a certain summary way about a subject. They act as symbols of the total information contained in a text or a body of texts, and they perform a function of identification. McPhee insists on the appropriateness of one clear label—*nonfiction*—when he says:

> Things that are cheap and tawdry in fiction work beautifully in nonfiction because they are true. That's why you should be careful not to abridge it, because it's the fundamental *power* you're dealing with. You *arrange it and present it*. There's lots of artistry. But you don't make it up. (emphasis added.)[5]

Without gainsaying the "power" of McPhee's truth, we can still insist on the gap between it and its image in the text. By referring again to Howarth's introduction to the *Reader,* we find an example of a more conventional critique of representation. Howarth engages the problem of attempting to define the character of a work by laws and labels as follows:

> He packs an impressive bag of narrative tricks, yet everyone calls his work "non-fiction." This label is frustrating, for it says not what a

book is but what it is *not*. Since "fiction" is presumably made up, imaginative, clever, and resourceful, a book of "non-fiction" must *not* be any of those things, perhaps not even a work of art. If the point seems a mere quibble over terms, try reversing the tables: are Faulkner's books on Mississippi "non-history" just because they are novels?[6]

The question of naming is crucial here. When McPhee talks rather ominously about the "power" of "nonfiction," and when Howarth discusses the misappropriation of the generic label, my sense is that they are treating narrative as a flexible category, the power or "tawdriness" of which can depend on what it is called. While both McPhee and Howarth might question the value of judging works according to strict conformity with generic codes, both would agree that in order to maintain the integrity of the subject, one must follow certain established journalistic practices.

Since McPhee's style has typically been praised for its dramatic scenic quality and its almost visceral presence, we might pose a more visual metaphor for the true replica. In Hollywood parlance, a photographic image is "true" inasmuch as it represents the view in the frame. But the replication accrues meaning in a totalized "image system" that defines the value of a particular frame or angle. The single photo has a certain correspondence to reality but only gains coherence in the total system of the film. The photolike realism of McPhee's writing scenes, by analogy, tends to divert any uncertainty about the authenticity of the subject. The match between image and representation tends to be regarded as natural and true to the extent that the film offers a coherent sum of images.

If we return to "A Sense of Where You Are," the McPhee story under consideration, we discover that much of the pleasure of reading it, and much of its coherency, comes through the play on the central metaphor of place. A sense of belonging where you are is crucial; without knowledge of place Bradley's game is unplayable. The *place* of place, however, is difficult to determine because it continually reactivates the chance for movement. Movement is possible in this game (of basketball and of writing) because one is aware of position. The concept of place is unthinkable without its supplement, movement.

For McPhee, and for the hero of his story, a sense of where you are determines the quality of play and openness, the degree of choice among nearly unlimited options. This developed sixth sense also operates by closing down options that break the rules or threaten to violate the logic or reason of the game. The sense of place makes sense only when there is a consensus, an agreement to play by the rules. It is ultimately hierarchical. To know one's place is to be aware of the possible positions in the sytem of placement. Knowing where you are requires the appropriate valuation of proportion, dignity, and self-confidence in the context of the game. It requires the knowledge of movement and the possible exchanges of place that can occur within the parameters of the game. It is in this context that we must begin to read this story.

The title, "A Sense of Where You Are," inspires a certain reliability and establishes propriety. It seems a logical starting place for a story on basketball, a game of placement. In this reading of Bill Bradley as a master of the game, all the "facts" point back to McPhee's title. Bradley's character is a remarkable one inasmuch as he has control of all possible actions within his given sphere. McPhee sets him apart from the other players because he senses that Bradley's superior ability lies in his willingness to play according to the "fundamental pattern" of the game.

An obvious connection between game and genre should be made here on the basis of the order internal to both. The concepts *game* and *genre* are imagined as sets of rules outside of which one can no longer play or work without fear of infraction, confusion, or the introduction of a new order of play. Play, which is vital to games, can never be freed in this sense. The writing model and the game plan must be executed within certain limited spheres and agreed-upon boundaries. Professionalism demands this, as does reason. When *free* play becomes a dimension of the game, there is little possibility that play can continue. In fact, play will be stopped and a forfeit imposed in order to regain game integrity and to protect against the further violence of rule-breaking. Literary journalism has shown that when genre opens to incorporate play, writing forms become subject to new scrutiny and a new set of rules that will attempt to contain the effects of the border crossing. Because a certain conception of historical truth is at stake when this crossing happens in journalism, the rules for writing "objectively" are systematically enforced.

The border crossing points to the very feature of genre that gives it the constant capacity for crossing over to be something it is not. Genre, like place, is both itself and not-itself at the same time, a paradoxical logic articulated most rigorously in "poststructuralist" thinking. This philosophy attempts to explain how genre, a structural concept that determines order, law, and placement, manifests doubt about the very possibility of certainty, fixity, and meaning. Law, or genre, then, introduces its own supplement. We might apply this reasoning to literary journalism by regarding the "novelistic" as a supplement to the "nonfiction." In a sense, the novel is added to the journalistic account. This addition points to a lack in the original text. It occupies the space where the "missing words" should or might have been. The novel is also a substitute for the factual report, what originally appeared in the text, since it directly challenges the fullness of that text's account. The supplement of fictional techniques in journalistic writing both makes up for and reveals a lack. The supplement, then, can be regarded as a feature of genre that is always possible.

We might draw another parallel between Bradley's basketball game and McPhee's writing, both games of pivots, pics, set-ups, and rebounds—games of will and containment. The components of Bradley's play adapt peculiarly well to McPhee's. McPhee ascribes various key qualities to Bradley: vision, discipline, concentration, freeing oneself up for the big play, understanding the "geometry of action," honing the "hunting" instinct, knowing the terrain, and manipulating the balance of power. All of these components might be made to fit a McPhee composite.[7] In fact, they might be made to fit many a "master narrative" of heroism. The play in both games is made to appear heroic, and the definition of heroism continually turns on mastering the logic of place, and the paradox of play.

Having a sense of where you are, learning what to do and then doing it as well as or better than your original models, is the hallmark of Bradley's character. Bradley's sense of place is nowhere more convincingly dramatized than in McPhee's replaying of a certain practice session on foreign turf:

Last summer, the floor of the Princeton gym was being resurfaced, so Bradley had to put in several practice sessions at the Lawrenceville School. His first afternoon at Lawrenceville, he began by shooting fourteen-foot jump shots from the right side. He got off to a bad start, and he kept missing them. Six in a row hit the back rim of the basket and bounced out. He stopped, looking discomfited, and seemed to be making an adjustment in his mind. Then he went up for another jump shot from the same spot and hit it cleanly. Four more shots went in without a miss, and then he paused and said, "You want to know something? That basket is about an inch and a half low." Some weeks later, I went back to Lawrenceville with a steel tape, borrowed a stepladder, and measured the height of the basket. It was nine feet ten and seven-eighths inches above the floor, or one and one-eighth inches too low.[8]

This remarkable sixth sense bestows a high degree of infallibility on Bradley that the other players do not have. Bradley is so sure of himself and his game that he is able to construe the difference of an inch and a half from a distance of fourteen feet. McPhee is also performing here. He has set up what rhetorician Michael Jordan calls a *situation structure* to combine the effects of subjective assessment and verifiable information.[9] He is also able to assess from his journalistic distance, a distance he foreshortens by his perfect knowledge of the writing game and his capacity to find and manipulate the "fundamental pattern."

Neither objective data nor opinion alone suffices to satisfy McPhee's desire for complete coverage of a situation or subject, so he provides both. But by linking point of view so intimately with observed data, and thereby rendering a type of "proof," McPhee avoids inviting the kind of scrutiny that accompanies the literary journalist's writing adventure when point of view and perspective become "too" subjective. This strategy works continuously throughout McPhee's piece. It is repeated as a kind of balancing act, an inconspicuous weighing and meting out of perspective and fact so that the thing itself seems to supply narrative structure and value.

Another example might serve to elaborate my point that McPhee's style takes on resemblance to truth in direct proportion to his ability to exemplify the logic of his theme: a sense of movement in place. Throughout the story, McPhee's prose illustrates an ability to allow Bradley to *speak for himself,* matching the rhythm of the writing with the motion and fluidity of Bradley's style. The need to explain the elegance of Bradley's strategy becomes less urgent if the writing itself reflects the movement of Bradley's "dance," a repeated metaphor for his game. Clearly nature is being improved upon by artifice. Like Bradley, McPhee always sets up for the next move. McPhee's ability to structure the narrative hinges on his faith in the importance and power of his performance to create meaning. Just as Bradley moves "for motion's sake, making plans and abandoning them" just to keep the basketball narrative going, giving it structure and place, so, too, McPhee moves around his subject angling first this way and then that to provide the most spectacular view of his subject in an associative and highly structured system of value.[10] McPhee creates a portrait of Bradley that takes shape during a skilled and close observation of him. That picture is also shaped according to the coherency of words and images to the overall writing plan, the "fundamental pattern" of which is to provide depth and balance.

In the following passage, McPhee sets out to substantiate his own ideas about Bradley's "most remarkable natural gift . . . his vision . . ."

> During a game, Bradley's eyes are always a glaze of panoptic atten-
> tion, for a basketball player needs to look at everything, focussing
> on nothing, until the last moment of commitment . . . Bradley's
> eyes close normally enough, but his astounding passes to teammates
> have given him, too, a reputation for being able to see out of the
> back of his head. To discover whether there was anything to all the
> claims for basketball players' peripheral vision, I asked Bradley to
> go with me to the office of Dr. Henry Abrams, a Princeton ophthal-
> mologist, who had agreed to measure Bradley's total field.[11]

The doctor discovered that Bradley had an abnormally large pe-
ripheral field of vision. McPhee also discovers, after further investigation, that Bradley probably affected his own field of vision by
practicing looking at objects out of the corners of his eyes as a kid.
The findings corroborate McPhee's initial hunch and fit in well
with the specific motif that Bradley's talent is as much acquired as
natural.

What is more interesting about this example stylistically is the way fact follows opinion. McPhee's observations of Bradley's eyes, which during a game "are always a glaze of panoptic attention," lead him to suggest the visit to the doctor's office. It appears that these astute *observations* have led to the "discovery" of the *new fact* of Bradley's supernormal field of sight, without which the comment that Bradley "can read the defense as if he were reading Braille" would be merely a pretty metaphor.[12] The focus is on verification; only after the trip to the doctor, and only after having all the collected material and notes at his disposal, does McPhee construct this specific image system. This point is worth noting because it emphasizes the importance of balance I have been talking about, between the real in literary play and the novelistic techniques in structuring the real. At the same time it makes the play— here the positioning of certain events in a certain sequence—seem less a product of a personal narrative imposition than a fitting cast to an identifiable subject or an external referent.

McPhee constantly goes back to the scene in order to verify the truth; it is as if without the verification we might not appreciate the value of Bradley's performance or the reliability of McPhee's word. What must be noted here is the perhaps obvious implication that nothing unsettles or mystifies if it can be structured and if that structure partakes of reality itself, not as the manufacturing of a metaphor for the real, but as a vision of the real—the empirically verifiable—as fundamentally structured. Logic, order, and meaning are the by-products of this structure. A hierarchy of value can be assigned around an implied center. This structure is the ground for the ideology of objectivity in journalism, and McPhee toes the line by preserving the center. He demystifies the story by certifying the facts. But despite McPhee's insistence on the power of truth, the way in which representation occurs always depends on artifice. The author disguises himself as recorder in order to temper the mediation between fact and story and to promote the "real illusion" that structure itself provides a natural and absolute system of identification rather than a true replica that is produced in the midst of narrative adventure.

I would like to briefly compare the work of another literary journalist, at least as adept at journalistic sleight of hand as McPhee, to show how craftily, if quietly, McPhee's prose operates on a story. Like Tom Wolfe, McPhee acknowledges the artifice of fictional technique, but more covertly. He conceives of the fictional part as a kind of glossy finish on the printed image. When we see Bradley "leaping like a salmon" or throwing a ball "like a pinch of salt" into the basket without looking behind him, we are faintly cognizant of authorial license, precisely because these images stand out from the "factual voice" that has previously been established. These particular images are not arbitrary; they, too, seem almost factual, fitting the fact of Bradley's leg strength and his supernormal range of vision, respectively. The secret of both Bradley's and McPhee's total game plans, however, remains mysteriously undisclosed.

In most of McPhee's work, the language sounds confident and playful but not sarcastic or hyperbolic. It is measured and polite, not extreme, and it is comprehensive but includes few stage bows, the summary and attention-getting authorial gestures that mark Wolfe's prose. Wolfe and McPhee use voice in opposite and symmetrical ways to produce the same rhetorical effects. If we compare two passages, the first from Wolfe's *Radical Chic* and the second from another McPhee story entitled *Encounters with the Archdruid,* we find a similar play on voice and point of view resulting from a very different writing style. In the following excerpt, Wolfe clearly speaks *for* an impersonal subject—a subject, nonetheless, with an idiom all its own. Because Wolfe needs to create a context and a story for the subject, a subject that has no access to Wolfe's overarching plan and cannot fully or effectively grasp itself within that context, he must control the selection and arrangement of information. In this particular instance, Wolfe's narrative design must fit the theme of social unease that results when radical meets chic or "black rage" and "white guilt" celebrate their hip collision. The effect, however, is of an intimate, very personal voice speaking for itself, without a script:

> Cheray tells her: "I've never met a Panther—this is a first for me!" . . . never dreaming that within forty-eight hours her words will be on the desk of the President of the United States . . .
>
> *This is a first for me.* But she is not alone in her thrill as the Black Panthers come trucking on in, into Lenny's house, Robert Bay, Don Cox the Panthers' Field Marshal from Oakland, Henry Miller the Harlem Panther defense captain, the Panther women—Christ, if the

Panthers don't know how to get it all together, as they say, the tight pants, the tight black turtlenecks, the leather coats, Cuban shades, Afros. But real Afros, not the ones that have been shaped and trimmed like a topiary hedge and sprayed until they have a sheen like acrylic wall-to-wall—but like funky, natural, scraggly . . . wild . . .
These are no civil-rights Negroes *wearing gray suits three sizes too big*—[13]

Wolfe's positioning of himself above and around the subject allows him to make the comparison between an Afro and a topiary hedge. Clearly, Wolfe needs to stand in for the subject in his own interests, and in the interests of the story. His assumption is that by doing this, by adopting the language of the subject yet maintaining distance, he can provide fuller and more complete coverage of the event.

In the McPhee excerpt that follows, the subject also only *seems* to speak for itself. McPhee is obviously allegorizing the conflict in order to dramatize the first encounter between the conservationist and former Sierra Club director, David Brower (the Archdruid), and the dam builder, United States Commissioner of Reclamation Floyd Dominy. At the same time he also appears to be having a little fun at Brower's expense. Still, he does not want to appear to preempt the story with the flashiness of his own associations so he starts off the section by attributing his analogy to a particular view and by representing that view in the third-person. The point of view is calculated to create a sense of the author's having researched the conflict carefully enough so that the metaphor develops as a natural analogue to the conflict:

In the view of conservationists, there is something special about dams, something—as conservation problems go—that is disproportionately and metaphysically sinister. The outermost circle of the Devil's world seems to be a moat filled mainly with DDT. Next to it is a moat of burning gasoline. Within that is a ring of pinheads each covered with a million people—and so on past phalanxed bulldozers and bicuspid chain saws into the absolute epicenter of Hell on earth, where stands a dam.[14]

McPhee is careful to mention right away that his comments represent the view of the conservationist since such a disclaimer automatically registers at a level of objectivity and balance that appears to be missing from the previously quoted Wolfe piece. Bulldozers and bicuspid chainsaws are not necessarily more "objective" elements in a story that dramatizes the conflict of development versus conservation than are tight pants, leather jackets, and Cuban shades in a story on the new black power. But the voices that render those ingredients appear to have a different relationship to the narrative. While Wolfe appears to be intimately involved, McPhee seems distanced and apart. In fact, one might say that Wolfe is habitually speaking *through* his subjects, using them as media, whereas McPhee is the medium through which his subjects speak.

Clearly, however, any neutral pose is illusory, and intimacy with or distance from the subject is impossible to measure. Both McPhee and Wolfe manipulate the voice of the subject to legitimize their own authorial acts and to give credibility to the story line, just as the story line must, since the genre demands it, seem to take form from the subject's voice. The literary journalists' assumption that the voice of the subject is a natural force, as opposed to writing, which supposedly has no emotional impact of its own, helps us to understand how writers depend on the reciprocity of world and word and how voice is used in narrative as a representation of ideal form or natural law.

In most journalistic writing, the personal mark of story arrangements obtrudes as little as possible on the record. What Hayden White calls history's "narrativization" appears to follow the natural order of the subject:

> Since its invention by Herodotus, traditional historiography has featured predominantly the belief that history itself consists of a congeries of lived stories, individual and collective, and that the principal task of historians is to uncover these stories and to retell them in a narrative, the truth of which would reside in the correspondence of the story told to the story lived by real people in the past. Thus conceived, the literary aspect of the historical narrative was supposed to inhere solely in certain stylistic embellishments that rendered the account vivid and interesting to the reader rather than in the kind of poetic inventiveness presumed to be characteristic of . . . fictional narratives.[15]

When the voice of the subject speaks *through* McPhee, when he makes himself "transparent," the law of journalism's genre, (i.e., of objectivity) acts as a naturalizing agent. The subject's voice is taken from its own presumed raw nature and recontextualized so that the objective account can penetrate and balance the emotional authenticity of the originating source. This complementarity is both meaningful and illusory: It is not *natural* itself, but is rather a matter of convention, by which representation is established as a legitimate means of assuring some immediate apprehension of the world. The referencing, therefore, of literary journalism in terms of an organic correspondence, a natural fit between world and story, protects authorial license in acts of representation. It also wards off the inclination to examine the notion of the "natural" and writing's function of denaturalization.

McPhee is a kind of latter-day Herodotus in that he seems to embellish without fictionalizing. He practices a style at once scrupulously accurate and boldly participatory and metaphorical, a style that has influenced successful "second generation" literary journalists like Mark Kramer and Tracy Kidder, and, one easily imagines, countless eager journalism students learning the literary craft.[16] He is not taught in courses on fiction and the American novel as are other nonfiction writers like Joan Didion, Norman Mailer, or Hunter Thompson, but in terms of my reading of the fictional act, where imagination, world, and text overlap, he leans into the "fictional space" of the novel, although less consciously, as much as any of the writers mentioned earlier.

Returning to the story at hand, we see, in fact, that McPhee sets Bradley up, as surely as Bradley sets up for a jump shot. He begins the story with a "curious event" that occurs before a play-off game during Bradley's senior year at Princeton:

> The game was played in Philadelphia and was the last of a tripleheader. The people were worn out, because most of them were emotionally committed to either Villanova or Temple—two local teams that had just been involved in enervating battles with Providence and Connecticut, respectively. . . . A group of Princeton boys shooting basketballs miscellaneously in preparation for still another game hardly promised to be a high point of the evening, but Bradley, whose routine in the warmup time is a gradual crescendo of activity, is more interesting to watch before a game than most players are in play.[17]

The rest of the paragraph describes in dramatic sequence Bradley's performance of "expandingly difficult jump shots" that go "cleanly through the basket with so few exceptions that the crowd began to murmur." Finally, after a series of "whirling reverse moves," more "jumpers," and sweeping "hook shots into the air" he begins to move

> in a semicircle around the court. First with his right hand, then with his left, he tried seven of these long, graceful shots—the most difficult ones in the orthodoxy of basketball—and ambidextrously made them all. The game had not even begun, but the presumably unimpressible Philadelphians were applauding like an audience at an opera.[18]

The crescendo of the play is at least matched by the crescendo of the written play as it moves dramatically to recreate the movement of the crowd's involvement from bored passivity to appreciative engagement. The comparison between the game and the opera further dramatizes the action on the court, and it works so well to evoke the requisite response of suspense and release that we forget it is McPhee's *own* assumption that the Philadelphians are "unimpressible."

McPhee offers a considered dramatization of the material, and his artistry informs the facts, just as his careful attention to detail earns for him (as it does for Wolfe) the poetic license to identify the subject through his metaphorical treatment of it. It allows him to be able to claim, for example, that "basketball is a hunting game . . . a player on offense either is standing around recovering his breath or is on the move, foxlike, looking for openings, sizing up chances . . ."[19] In McPhee's and Wolfe's careful attention to the reconstruction of the voices of their subjects we never hear the nostalgic or self-conscious voice of the author who cannot quite reclaim the subject, as we do, for example, in Sara Davidson's, Norman Mailer's, or Joan Didion's work. Instead, they both rely on the subject's "natural" voice to make the limitations of writing and the work of the author seem to disappear.

An important question for the practice of journalism, then, is how, when a writer takes up a position that claims to be balanced and neutral, can he or she make value judgments at the same time? One way of examining this question is to regard the recorded event not only as a narrative recreation but as a version of history that was deemed apt and legitimate enough to record in the first place. Howarth's introduction to "Encounters with the Archdruid," which appears in *The John McPhee Reader,* is, in this context, an interesting prelude in that he tells us McPhee had already planned the book before finding his protagonist. The plan, he writes, "was a bit formulaic; it resembled Boswell's jostling of Dr. Johnson into conversations of quotable prose." However, he defends the use of this formula, an excellent example in at least one respect that McPhee is absolutely impartial "on the issues he dramatizes."[20] It is worthwhile quoting the rest of Howarth's short commentary on the piece because, even as he underscores McPhee's set up of the balanced equation, he demonstrates how an uncritical notion of balance can be made to serve impartiality:

> For every point Brower scores on the beauties of wildness, his opponents respond with sensible defenses of progress. . . . Brower is no mere Druid, a worshipper of trees, nor are his adversaries simply out to exploit the land. . . . The story exemplifies how facts lend themselves to McPhee's imaginative handling; the lake, river, dam, and raft become his emblems of rigidity or flexibility, expressing a scale of values without forcing him to "take a position" on these controversial issues.[21]

We may be able to agree with Howarth about the surface effects of the story: it provides balanced views, the characters are not one-dimensional (good or bad subjects), and the landscape seems to work as symbol of the conflict being recounted. There is something odd, however, in the claim that McPhee can maintain a creditable *distance* between expressing value and taking a position. In fact, all the way through the story McPhee is telling us how to respond by giving either Dominy or Brower the last word, by making one seem heroic in one instance, a coward or a fool the next. While McPhee may not be assigning an absolute value either to irrigation or to the preservation of the natural form of the river, he is clearly taking the position that in relative terms, *nature*—in this case the Colorado River—must ultimately be allowed to speak for itself. In this story, McPhee wants to convey the idea that balance resides in nature and so, too, in the narrative. Of course, the story has been selected according to a specific criterion of conflict that seems to call for arbitration, which, as always, is the task of the author. Moreover, since nature in this case has no clear voice of its own—one might say it is forcefully mute—McPhee must provide it with one.

The way in which the news is selected for presentation is the subject of several studies on media practice. The importance of the "story" feature, how much drama the event can promise to news presentation, is clearly high on the list of necessary characteristics. As Hayden White writes, it is the drama of the story "line" that reinvests reality with truth:

> The authority of the historical narrative is the authority of reality itself; the historical account endows this reality with form and thereby makes it desirable by the imposition upon its processes of the formal coherency that only stories possess.[22]

The position of neutrality becomes more difficult to imagine when we understand that selection of story material hinges on specific cultural practices that may not be flexible enough to accommodate events that do not fit the routine. "Events," White continues,

> are real not because they occurred but because, first, they were remembered and, second, they are capable of finding a place in a chronologically ordered sequence. . . . Unless at least two versions of the same set of events can be imagined, there is no reason for the historian to take upon himself the authority of giving the true account.[23]

Part of the problem in claiming neutrality, then, is that the fiction–nonfiction categories do not remain separate despite the insistent implication of objectivity in the nonfictional label. McPhee claims a fundamental "power" for the historical real, but not until the real is packaged and reported and charged with effects does the event become part of our narrative history, part of our structure of meaning.

If we agree, therefore, with White and with Barthes that history somehow postdates its own packaging, then we can begin to measure the effects of the linguistic spectacle on journalism. The *label* "literary journalism" has the capacity to force a new way of reading news as an adventure of language. It does not make changes *in* the text. The text itself is plural—we may believe it if it is labeled nonfiction or we may suspend belief if it is labeled novel; one story contains the difference within itself to allow completely varied readings, at the instigation of a word. But the label has helped refocus attention on the role of narrative in determining news value and allowed a more complete and sophisticated understanding of the way in which a story takes on meaning.

Although the role that various narrative strategies play in "uncovering" history is not underplayed at the surface level of the text in McPhee's work, the assumption of categorical differences between nonfictional and fictional discourses remains intact. McPhee regards fiction as an interfering or distorting mechanism that endangers the process of the recovery work, and he dismisses the notion that nonfiction and fiction have common aspects (their substitution of language for events) as secondary philosophical concerns, unrelated to the business of reproducing objective accounts of reality. For McPhee, the distinction between fact-gathering in the field and invention on the page is predicated, as in Wolfe's work, on the accepted notion of objectivity; it is the quest for knowledge itself that is at stake when two genres collide. I refer again to White:

> Myths and the ideologies based on them presuppose the adequacy of stories to the representation of the reality whose meaning they purport to reveal. When belief in this adequacy begins to wane, the entire cultural edifice of a society enters into crisis, because not only is a specific system of beliefs undermined but the very condition of possibility of socially significant belief is eroded.[24]

For journalists, the possibility that an objective narrative devised precisely for the purpose of preserving reality in a recognized and shared body of knowledge is working at the service of a "myth" or an "ideology" is already to approach the state of crisis. In order to preserve the metaphors that lend credibility to the discovery of real knowledge, and the "condition of possibility" for emerging truth, the author needs a verifiable, empirical world in which to insert himself. In McPhee's case, there is a kind of fearlessness in the prose that is achieved by the easy conjunction of fact and metaphor. He does not confront the interesting contradiction that truth can be mediated by the "arrangement and presentation" and that the "power" of truth is a function of its representation in the text; in other words, that truth and representation reflect one another but do not necessarily converge.

This is an especially vital point in regard to McPhee since his work has the character of an equation. As I suggested earlier, he balances the need for subjective assessment of his subject (his metaphorical treatment of it) with the objective field of collected data. In the process, language, the structure of the equation, becomes the instrument by which reportorial objectivity, the cornerstone of generic law for journalism, is both reasserted and dismantled. The paradoxical position for McPhee is that he both desires the power of narrative to convey the real and resists its continual and infinite power to distort it. White writes:

> Narrative becomes a problem only when we wish to give to real events the form of story. It is because real events do not offer themselves as stories that their narrativization is so difficult.[25]

Without structure, without a "moral" order of meaning, events look like mere compilations of information, lists that have no connection with human intercourse. In McPhee's meticulously crafted writing, which appears so painstakingly true to life and which posits a natural correspondence between subject and style, the contradiction nevertheless remains: the subject is clearly *fashioned* and at the same time supposedly undisturbed in its essential nature. There is a clear irony in the configuration. The structure and codes of McPhee's writing resist denaturalization precisely because of his unwillingness to acknowledge the narrative invasion on an "objective" subject.

History unfolds as an interpretive and performative gesture of recovery. It is personal and provisional, and in that sense objective. In the balance McPhee attempts to preserve, there is no security of objectivity but rather a reconfirmation of the determining, subjective feature of the law of narrative that presents truth by imposing order. Because McPhee is both spectator of the real and manipulator or player in its representation, he is in a position to produce and stage the heroism of his subject that is both Bradley's game and his own game of literary journalism. The assumed value of the fundamental pattern of the court play is the same one that seems to characterize McPhee's writing activity. Perhaps it is this very neat formula itself that remains still to be reevaluated, for it may be only those who, like McPhee, clearly operate according to the fundamental pattern of journalism who can most easily, if unwittingly, twist its logic and stretch its rules.

The operative assumption in literary journalism is that if we desire the certainty of order we must recognize that we gain it at the price of its mediation through systems of meaning designed to control the reading process. For McPhee, that understanding is both the given from which he begins to write and the burden he refuses fully to disclose. What then becomes of objectivity when it is only the *sense* of where you are (the sixth and nonverifiable sense) that is translatable into language, already a metaphor for the place of the absent thing? What if this place is also a "non-place," the point that is always in motion? McPhee's fictionalizing act, whether he acknowledges it or not, is grounded in the same logic as all narrative, one that seeks, through illusion, a perfect apprehension of the world. In stubbornly resisting the fiction label, McPhee comes no closer to the world of fact. On the contrary, his work only reaffirms the need for literary journalists to maintain a certain play between the verifiable and the fictional, which invests the facts with a common value. He participates in the fictionalizing act because he takes up this illusory position of balance when he attempts to narrow the gap between narrative versions of history and the definitive real, and because he so completely forces his metaphors to conform to a "naturalistic" reading of the subject, as if they belonged outside of writing.

John McPhee *reinforces* the importance of a feature of his method that, at the same time, he seeks to *reform:* a sense of privilege without involvement. This position creates moral as well as aesthetic tension. How does one remain uninvolved in a position of privilege? How does one remain distant and close simultaneously? What effects does the dilemma of privileged noninvolvement have on the story and the story sources? On the writing of journalism and the genre itself?

McPhee obviously uses the nonfiction label in an attempt to tip the scales and to underplay the fictional elements of the book. The equivalence between the aesthetic use of label and its appropriateness to truth is partly achieved—and we are reminded of this quite graphically in Howarth's introductory comments to "Encounters with the Archdruid"—by an imposition on the material that has been composed with the label in mind. In imagining the breakdown of genre and the dysfunctional and misleading fact–fiction dichotomy, the possibility of the third story emerges—one that is not necessarily a synthesis of the two (the novel and the report), but is instead an expression of the ambiguous in and out movement as the necessary movement of writing itself. With McPhee, the movement remains unconsious in the text, and the label sticks as a sign of the story's grounding in one representational mode. The balance he affects, however, the fictional poise of it, remains to negotiate and color and finally to narrate the facts.

This is not to say that facts are not facts. It is important that McPhee retain the label nonfiction so that the subject of his narrative can produce the power that he speaks of, the power of legitimacy conferred by the label that promises a true story. By now, though, it should be clear that the object of this chapter is to examine the true story as a contradiction in terms that the rethinking of journalism as literary has helped to begin. For McPhee, the struggle to make and assign order to the world out there and the replication of that ordered world in the text meet in the writing process. The novelistic design of the text affords McPhee the satisfaction of finding the one "best place" for the facts, by placing them inexorably at the mercy of invention.

NOTES

1. William Dean Howells, *The Rise of Silas Lapham* (New York: Random House, 1951), p. 5.

2. See Ronald Weber, *The Literature of Fact* (Athens: Ohio University Press, 1981). I take Weber's comments on McPhee to typify certain New Critical ideas on the organic unity of form and content, or style and meaning. He writes, for example, that McPhee has the capacity to "make factual experience meaningful through the process of giving weight and significance to the particular, of endowing the particular with resonant meanings," p. 116.

3. William Howarth, introduction to *The John McPhee Reader,* William Howarth, ed. (New York: Vintage Books, 1977), p. viii.

4. Roland Barthes, "Introduction to the Structural Analysis of Narratives," in *Image, Music, Text,* Stephen Heath, trans. (New York: Hill and Wang, 1977), p. 124.

5. Quote attributed to John McPhee; see *The Literary Journalists,* Norman Sims, ed. (New York: Ballantine Books, 1984), p. 3.

6. Howarth, *McPhee Reader,* p. vii.

7. John McPhee, "A Sense of Where You Are," in *The John McPhee Reader,* pp. 3–21.

8. ibid., p. 7.

9. Michael P. Jordan, *Rhetoric of Everyday English Texts* (London: Allen & Unwin, 1984), p. 89.

10. McPhee, "A Sense of Where You Are," pp. 16–17.

11. ibid., p. 19.

12. ibid., p. 20.

13. Tom Wolfe, *Radical Chic and Mau-Mauing the Flak Catchers* (New York: Bantam, 1970), p. 8.

14. McPhee, *Encounters with the Archdruid* (New York: Farrar, Straus and Giroux, 1971), p. 158.

15. Hayden White, introduction to *The Content of the Form: Narrative Discourse and Historical Representation* (Baltimore: Johns Hopkins University Press, 1987), pps. ix–x.

16. See Sims, *The Literary Journalists,* a collection of interviews with and stories by literary journalists. "Second generation" literary journalists discuss the influence of writers like McPhee on their work.

17. McPhee, "A Sense of Where You Are," p. 4.

18. ibid., p. 4.

19. ibid., p. 16.

20. Howarth, *McPhee Reader,* p. 189.

21. ibid., pp. 189–90.

22. White, *The Content of the Form,* p. 20.

23. ibid., p. 20.

24. ibid., p. x.

25. ibid., p. 5.

PART III

10

Artists in Uniform

Mary McCarthy

March, 1953

The Colonel went out sailing,
He spoke with Turk and Jew . . .

"Pour it on, Colonel," cried the young man in the Dacron suit excitedly, making his first sortie into the club-car conversation. His face was white as Roquefort and of a glistening, cheeselike texture; he had a shock of tow-colored hair, badly cut and greasy, and a snub nose with large gray pores. Under his darting eyes were two black craters. He appeared to be under some intense nervous strain and had sat the night before in the club car drinking bourbon with beer chasers and leafing through magazines which he frowningly tossed aside, like cards into a discard heap. This morning he had come in late, with a hangdog, hangover look, and had been sitting tensely forward on a settee, smoking cigarettes and following the conversation with little twitches of the nose and quivers of the body, as a dog follows a human conversation, veering its mistrustful eyeballs from one speaker to another and raising its head eagerly at its master's voice. The colonel's voice, rich and light and plausible, had in fact abruptly risen and swollen, as he pronounced his last sentence. "I can tell you one thing," he had said harshly. "They weren't named Ryan or Murphy!"

A sort of sigh, as of consummation, ran through the club car. "Pour it on, Colonel, give it to them, Colonel, that's right, Colonel," urged the young man in a transport of admiration. The colonel fingered his collar and modestly smiled. He was a thin, hawklike, black-haired handsome man with a bright blue bloodshot eye and a well-pressed, well-tailored uniform that did not show the effects of the heat—the train, westbound for St. Louis, was passing through Indiana, and, as usual in a heat wave, the airconditioning had not met the test. He wore the Air Force insignia, and there was something in his light-boned, spruce figure and keen, knifelike profile that suggested a classic image of the aviator, ready to cut, piercing, into space. In base fact, however, the colonel was in procurement, as we heard him tell the mining engineer who had just bought him a drink. From several silken hints that parachuted into the talk, it was patent to us that the colonel was a man who knew how to enjoy this earth and its pleasures: he led, he gave us to think, a bachelor's life of abstemious dissipation and well-rounded sensuality. He had accepted the engineer's drink with a mere nod of the glass in acknowledgement, like a genial Mars quaffing a libation; there was clearly no prospect of his buying a second in return, not if the train were to travel from here to the Mojave Desert. In the same way, an understanding had arisen that I, the only woman in the club car, had become the colonel's perquisite; it was taken for granted, without an invitation's being issued, that I was to lunch with him in St. Louis, where we each had a wait between trains—my plans for seeing the city in a taxicab were dished.

From the beginning, as we eyed each other over my volume of Dickens ("*The Christmas Carol?*" suggested the colonel, opening relations), I had guessed that the colonel was of Irish stock, and this, I felt, gave me an advantage, for he did not suspect the same of me; strangely so, for I am supposed to have the map of Ireland written on my features. In fact, he had just wagered, with a jaunty, sidelong grin at the mining engineer, that my people "came from Boston from way back," and that I—narrowed glance, running, like steel measuring-tape, up and down my form—was a professional sculptress. I might have laughed this off, as a crudely bad guess like his *Christmas Carol,* if I had not seen the engineer nodding gravely, like an idol, and the peculiar young man bobbing his head up and down in mute applause and agreement. I was wearing a bright apple-green raw silk blouse and a dark-green rather full raw silk skirt, plus a pair of pink glass earrings; my hair was done up in a bun. It came to me, for the first time, with a sort of dawning horror, that I had begun, in the course of years, without ever guessing it, to look irrevocably Bohemian. Refracted from the three men's eyes was a strange vision of myself as an artist, through and through, stained with my occupation like the dyer's hand. All I lacked, apparently, was a pair of sandals. My sick heart sank to my Ferragamo shoes; I had always particularly preened myself on being an artist in disguise. And it was not only a question of personal vanity—it seemed to me that the writer or intellectual had a certain missionary usefulness in just such accidental gatherings as this, if he spoke not as an intellectual but as a normal member of the public. Now, thanks to the colonel, I slowly became aware that my contributions to the club-car conversation were being watched and assessed as coming from *a certain quarter.* My costume, it seemed, carefully assembled as it had been at an expensive shop, was to these observers simply a uniform that blazoned a caste and allegiance just as plainly as the colonel's khaki and eagles. "*Gardez,*" I said to myself. But, as the conversation grew tenser and I endeavored to keep cool, I began to writhe within myself, and every time I looked down, my contrasting greens seemed to be growing more and more lurid and taking on an almost menacing light, like leaves just before a storm that lift their bright undersides as the air becomes darker. We had been speaking, of course, of Russia, and I had mentioned a study that had been made at Harvard of political attitudes among Iron Curtain refugees. Suddenly, the colonel had smiled. "They're pretty Red at Harvard, I'm given to understand," he observed in a comfortable tone, while the young man twitched and quivered urgently. The eyes of all the men settled on me and waited. I flushed as I saw myself reflected. The woodland greens of my dress were turning to their complementary red, like a color-experiment in psychology or a traffic light changing. Down at the other end of the club car, a man looked up from his paper. I pulled myself together. "Set your mind at rest, Colonel," I remarked dryly. "I know Harvard very well and they're conservative to the point of dullness. The only thing crimson is the football team." This disparagement had its effect. "So . . . ?" queried the colonel. "I thought there was some professor. . . ." I shook my head. "Absolutely not. There used to be a few fellow-travelers, but they're very quiet these days, when they haven't absolutely recanted. The general atmosphere is more anti-Communist than the Vatican." The colonel and the mining engineer exchanged a thoughtful stare and seemed to agree that the Delphic oracle that had just pronounced knew whereof it spoke. "Glad to hear it," said the colonel. The engineer frowned and shook his fat wattles; he was a stately, gray-haired, plump man with small hands and feet and the pampered, finical tidiness of a small-town widow. "There's so much hearsay these days," he exclaimed vexedly. "You don't know *what* to believe."

I reopened my book with an air of having closed the subject and read a paragraph three times over. I exulted to think that I had made a modest contribution to sanity in our times, and I imagined my words pyramiding like a chain letter—the colonel telling a fellow-officer on the veranda of a club in Texas, the engineer halting a works-superintendent in a Colorado mine shaft: "I met a woman on the train who claims . . . Yes, absolutely. . . ." Of course, I did not know Harvard as thoroughly as I pretended, but I forgave myself by thinking it was the convention of such club-car symposia in our positivistic country to speak from the horse's mouth.

Meanwhile, across the aisle, the engineer and the colonel continued their talk in slightly lowered voices. From time to time, the colonel's polished index-fingernail scratched his burnished black head and his knowing blue eye forayed occasionally toward me. I saw that still I was a doubtful quantity to them, a movement in the bushes, a noise, a flicker, that was figuring in the crenelated thought as "she." The subject of Reds in our colleges had not, alas, been finished; they were speaking now of another university and a woman faculty-member who had been issuing Communist statements. This story somehow, I thought angrily, had managed to appear in the newspapers without my knowledge, while these men were conversant with it; I recognized a big chink in the armor of my authority. Looking up from my book, I began to question them sharply, as though they were reporting some unheard-of natural phenomenon. "When?" I demanded. "Where did you see it? What was her name?" This request for the professor's name was a headlong attempt on my part to buttress my position, the implication being that the identities of all university professors were known to me and that if I were but given the name I could promptly clarify the matter. To admit that there was a single Communist in our academic system whose activities were hidden from me imperiled, I instinctively felt, all the small good I had done here. Moreover, in the back of my mind, I had a supreme confidence that these men were wrong: the story, I supposed, was some tattered piece of misinformation they had picked up from a gossip column. Pride, as usual, preceded my fall. To the colonel, the demand for the name was not specific but generic: what *kind* of name was the question he presumed me to be asking. "Oh," he said slowly with a luxurious yawn, "Finkelstein or Fishbein or Feinstein." He lolled back in his seat with a side glance at the engineer, who deeply nodded. There was a voluptuary pause, as the implication sank in. I bit my lip, regarding this as a mere diversionary tactic. "Please!" I said impatiently. "Can't you remember exactly?" The colonel shook his head and then his spare cheekbones suddenly reddened and he looked directly at me. "I can tell you one thing," he exclaimed irefully. "They weren't named Ryan or Murphy."

The colonel went no further; it was quite unnecessary. In an instant, the young man was at his side, yapping excitedly and actually picking at the military sleeve. The poor thing was transformed, like some creature in a fairy tale whom a magic word releases from silence. "That's right, Colonel," he happily repeated. "I know them. *I* was at Harvard in the business school, studying accountancy. I left, I couldn't take it." He threw a poisonous glance at me, and the colonel, who had been regarding him somewhat doubtfully, now put on an alert expression and inclined an ear for his confidences. The man at the other end of the car folded his newspaper solemnly and took a seat by the young man's side. "They're all Reds, Colonel," said the young man. "They teach it in the classroom. I came back here to Missouri. It made me sick to listen to the stuff they handed out. If you didn't hand it back, they flunked you. Don't let anybody tell you different." "You are wrong," I said coldly and closed my book and rose. The

young man was still talking eagerly, and the three men were lean-
ing forward to catch his every gasping word, like three astute detec-
tives over a dying informer, when I reached the door and cast a last
look over my shoulder at them. For an instant, the colonel's eye
met mine and I felt his scrutiny processing my green back as I
tugged open the door and met a blast of hot air, blowing my full
skirt wide. Behind me, in my fancy, I saw four sets of shrugging
brows.

In my own car, I sat down, opposite two fat nuns, and tried to
assemble my thoughts. I ought to have spoken, I felt, and yet,
what could I have said? It occurred to me that the four men had
perhaps not realized why I had left the club car with such abrupt-
ness: was it possible that they thought I was a Communist, who
feared to be unmasked? I spurned this possibility, and yet it made
me uneasy. For some reason, it troubled my *amour-propre* to think
of my anti-Communist self living on, so to speak, green in their
collective memory as a Communist or fellow-traveler. In fact,
though I did not give a fig for the men, I hated the idea, while a
few years ago I should have counted it a great joke. This, it seemed
to me, was a measure of the change in the social climate. I had
always scoffed at the notion of liberals "living in fear" of political
demagoguery in America, but now I had to admit that if I was not
fearful, I was at least uncomfortable in the supposition that anybody, anybody whatever, could think of me, precious me, as a
Communist. A remoter possibility was, of course, that back there
my departure was being ascribed to Jewishness, and this too annoyed me. I am in fact a quarter Jewish, and though I did not
"hate" the idea of being taken for a Jew, I did not precisely like it,
particularly under these circumstances. I wished it to be clear that I
had left the club car for intellectual and principled reasons; I
wanted those men to know that it was not I, but my principles, that
had been offended. To let them conjecture that I had left because I
was Jewish would imply that only a Jew could be affronted by an
anti-Semitic outburst; a terrible idea. Aside from anything else, it
voided the whole concept of transcendence, which was very close
to my heart, the concept that man is more than his circumstances,
more even than himself.

However you looked at the episode, I said to myself nervously, I had not acquitted myself well. I ought to have done or said something concrete and unmistakable. From this, I slid glassily to the thought that those men ought to be punished, the colonel, in particular, who occupied a responsible position. In a minute, I was framing a businesslike letter to the Chief of Staff, deploring the colonel's conduct as unbecoming to an officer and identifying him by rank and post, since unfortunately I did not know his name. Earlier in the conversation, he had passed some comments on "Harry" that bordered positively on treason, I said to myself triumphantly. A vivid image of the proceedings against him presented itself to my imagination: the long military tribunal with a row of stern soldierly faces glaring down at the colonel. I myself occupied only an inconspicuous corner of this tableau, for, to tell the truth, I did not relish the role of the witness. Perhaps it would be wiser to let the matter drop . . . ? We were nearing St. Louis now; the colonel had come back into my car, and the young accountant had followed him, still talking feverishly. I pretended not to see them and turned to the two nuns, as if for sanctuary from this world and its hatred and revenges. Out of the corner of my eye, I watched the colonel, who now looked wry and restless; he shrank against the window as the young man made a place for himself amid the colonel's smart luggage and continued to express his views in a pale breathless voice. I smiled to think that the colonel was paying the piper. For the colonel, anti-Semitism was simply an aspect of urbanity, like a knowledge of hotels or women. This frantic psychopath of an accountant was serving him as a nemesis, just as the German people had been served by their psychopath, Hitler. Colonel, I adjured him, you have chosen, between him and me; measure the depth of your error and make the best of it! No intervention on my part was now necessary; justice had been meted out. Nevertheless, my heart was still throbbing violently, as if I were on the verge of some dangerous action. What was I to do, I kept asking myself, as I chatted with the nuns, if the colonel were to hold me to that lunch? And I slowly and apprehensively revolved this question, just as though it were a matter of the most serious import. It seemed to me that if I did not lunch with him—and I had no intention of doing so—I had the dreadful obligation of telling him why.

He was waiting for me as I descended the car steps. "Aren't you coming to lunch with me?" he called out and moved up to take my elbow. I began to tremble with audacity. "No," I said firmly, picking up my suitcase and draping an olive-green linen duster over my arm. "I can't lunch with you." He quirked a wiry black eyebrow. "Why not?" he said. "I understood it was all arranged." He reached for my suitcase. "No," I said, holding on to the suitcase. "I can't." I took a deep breath. "I have to tell you. I think you should be *ashamed* of yourself, Colonel, for what you said in the club car." The colonel stared: I mechanically waved for a redcap, who took my bag and coat and went off. The colonel and I stood facing each other on the emptying platform. "What do you mean?" he inquired in a low, almost clandestine tone. "Those anti-Semitic remarks," I muttered, resolutely. "You ought to be *ashamed*." The colonel gave a quick, relieved laugh. "Oh, come now," he protested. "I'm sorry," I said. "I can't have lunch with anybody who feels that way about the Jews." The colonel put down his attaché case and scratched the back of his lean neck. "Oh, come now," he repeated, with a look of amusement. "You're not Jewish, are you?" "No," I said quickly. "Well, then . . ." said the colonel, spreading his hands in a gesture of bafflement. I saw that he was truly surprised and slightly hurt by my criticism, and this made me feel wretchedly embarrassed and even apologetic, on my side, as though I had called attention to some physical defect in him, of which he himself was unconscious. "But I might have been," I stammered. "You had no way of knowing. You oughtn't to talk like that." I recognized, too late, that I was strangely reducing the whole matter to a question of etiquette: "Don't start anti-Semitic talk before making sure there are no Jews present." "Oh, hell," said the colonel, easily. "I can tell a Jew." "No, you can't," I retorted, thinking of my Jewish grandmother, for by Nazi criteria I was Jewish. "Of course I can," he insisted. "So can you." We had begun to walk down the platform side by side, disputing with a restrained passion that isolated us like a pair of lovers. All at once, the colonel halted, as though struck with a thought. "What *are* you, anyway?" he said meditatively, regarding my dark hair, green blouse, and pink earrings. Inside myself, I began to laugh. "Oh," I said gaily, playing out the trump I had been saving. "I'm Irish, like you, Colonel." "How did you know?" he said amazedly. I laughed aloud. "I can tell an Irishman," I taunted. The colonel frowned.

"What's your family name?" he said brusquely. "McCarthy." He lifed an eyebrow, in defeat, and then quickly took note of my wedding ring. "That your maiden name?" I nodded. Under this peremptory questioning, I had the peculiar sensation that I get when I am lying; I began to feel that "McCarthy" was a nom de plume, a coinage of my artistic personality. But the colonel appeared to be satisfied. "Hell," he said, "come on to lunch, then. With a fine name like that, you and I should be friends." I still shook my head, though by this time we were pacing outside the station restaurant; my baggage had been checked in a locker; sweat was running down my face and I felt exhausted and hungry. I knew that I was weakening and I wanted only an excuse to yield and go inside with him. The colonel seemed to sense this. "Hell," he conceded. "You've got me wrong. I've nothing against the Jews. Back there in the club car, I was just stating a simple fact: you won't find an Irishman sounding off for the Commies. You can't deny that, can you?"

His voice rose persuasively; he took my arm. In the heat, I wilted and we went into the air-conditioned cocktail lounge. The colonel ordered two old-fashioneds. The room was dark as a cave and produced, in the midst of the hot midday, a hallucinated feeling, as though time had ceased, with the weather, and we were in eternity together. As the colonel prepared to relax, I made a tremendous effort to guide the conversation along rational, purposive lines; my only justification for being here would be to convert the colonel. "There *have* been Irishmen associated with the Communist party," I said suddenly, when the drinks came. "I can think of two." "Oh, hell," said the colonel, "every race and nation has its traitors. What I mean is, you won't find them in numbers. You've got to admit the Communists in this country are ninety percent Jewish." "But the Jews in this country aren't ninety percent Communist," I retorted.

As he stirred his drink, restively, I began to try to show him the reasons why the Communist movement in America had attracted such a large number, relatively, of Jews: how the Communists had been anti-Nazi when nobody else seemed to care what happened to the Jews in Germany; how the Communists still capitalized on a Jewish fear of fascism; how many Jews had become, after Buchenwald, traumatized by this fear. . . .

But the colonel was scarcely listening. An impatient frown
rested on his jaunty features. "I don't get it," he said slowly. "Why
should you be for them, with a name like yours?" "I'm *not* for the
Communists," I cried. "I'm just trying to explain to you—" "For
the Jews," the colonel interrupted, irritable now himself. "I've
heard of such people but I never met one before." "I'm not 'for'
them," I protested. "You don't understand. I'm not for *any* race or
nation. I'm against those who are against them." This word, *them,*
with a sort of slurring circle drawn round it, was beginning to
sound ugly to me. Automatically, in arguing with him, I seemed to
have slipped into the colonel's style of thought. It occurred to me
that defense of the Jews could be a subtle and safe form of anti-
Semitism, an exercise of patronage: as a rational Gentile, one
could feel superior both to the Jews and the anti-Semites. There
could be no doubt that the Jewish question evoked a curious
stealthy lust or concupiscence. I could feel it now vibrating between us over the dark table. If I had been a good person, I should
unquestionably have got up and left.

"I don't get it," repeated the colonel. "How were you brought
up? Were your people this way too?" It was manifest that an odd
reversal had taken place; each of us regarded the other as "abnormal" and was attempting to understand the etiology of a disease.
"Many of my people think just as you do," I said, smiling coldly.
"It seems to be a sickness to which the Irish are prone. Perhaps it's
due to the potato diet," I said sweetly, having divined that the
colonel came from a social stratum somewhat lower than my own.

But the colonel's hide was tough. "You've got me wrong," he
reiterated, with an almost plaintive laugh. "I don't dislike the Jews.
I've got a lot of Jewish friends. Among themselves, they think just
as I do, mark my words. I tell you what it is," he added rumina-
tively, with a thoughtful prod of his muddler, "I draw a distinction
between a kike and a Jew." I groaned. "Colonel, I've never heard
an anti-Semite who didn't draw that distinction. You know what
Otto Kahn said? 'A kike is a Jewish gentleman who has just left the
room.' " The colonel did not laugh. "I don't hold it against some of
them," he persisted, in a tone of pensive justice. "It's not their
fault if they were born that way. That's what I tell them, and they
respect me for my honesty. I've had a lot of discussions; in procurement, you have to do business with them, and the Jews are the first

to admit that you'll find more chiselers among their race than among the rest of mankind." "It's not a race," I interjected wearily, but the colonel pressed on. "If I deal with a Jewish manufacturer, I can't bank on his word. I've seen it again and again, every damned time. When I deal with a Gentile, I can trust him to make delivery as promised. That's the difference between the two races. They're just a different breed. They don't have standards of honesty, even among each other." I sighed, feeling unequal to arguing the colonel's personal experience.

"Look," I said, "you may be dealing with an industry where the Jewish manufacturers are the most recent comers and feel they have to cut corners to compete with the established firms. I've heard that said about Jewish cattle-dealers, who are supposed to be extra sharp. But what I think, really, is that you notice it when a Jewish firm fails to meet an agreement and don't notice it when it's a Yankee." "Hah," said the colonel. "They'll tell you what I'm telling you themselves, if you get to know them and go into their homes. You won't believe it, but some of my best friends are Jews," he said, simply and thoughtfully, with an air of originality. "They may be *your* best friends, Colonel," I retorted, "but you are not theirs. I defy you to tell me that you talk to them as you're talking now." "Sure," said the Colonel, easily. "More or less." "They must be very queer Jews you know," I observed tartly, and I began to wonder whether there indeed existed a peculiar class of Jews whose function in life was to be "friends" with such people as the colonel. It was difficult to think that all the anti-Semites who made the colonel's assertion were the victims of a cruel self-deception.

A dispirited silence followed. I was not one of those liberals who believed that the Jews, alone among peoples, possessed no characteristics whatever of a distinguishing nature—this would mean they had no history and no culture, a charge which should be leveled against them only by an anti-Semite. Certainly, types of Jews could be noted and patterns of Jewish thought and feeling: Jewish humor, Jewish rationality, and so on, not that every Jew reflected every attribute of Jewish life or history. But somehow, with the colonel, I dared not concede that there was such a thing as a Jew: I saw the sad meaning of the assertion that a Jew was a person whom other people thought was Jewish.

Hopeless, however, to convey this to the colonel. The desolate truth was that the colonel was extremely stupid, and it came to me, as we sat there, glumly ordering lunch, that for extremely stupid people anti-Semitism was a form of intellectuality, the sole form of intellectuality of which they were capable. It represented, in a rudimentary way, the ability to make categories, to generalize. Hence a thing I had noted before but never understood: the fact that anti-Semitic statements were generally delivered in an atmosphere of profundity. Furrowed brows attended these speculative distinctions between a kike and a Jew, these little empirical laws that you can't know one without knowing them all. To arrive, indeed, at the idea of a Jew was, for these grouping minds, an exercise in Platonic thought, a discovery of essence, and to be able to add the great corollary, "Some of my best friends are Jews," was to find the philosopher's cleft between essence and existence. From this, it would seem, followed the querulous obstinacy with which the anti-Semite clung to his concept; to be deprived of this intellectual tool by missionaries of tolerance would be, for persons like the colonel, the equivalent of Western man's losing the syllogism: a lapse into animal darkness. In the club car, we had just witnessed an example: the colonel with his anti-Semitic observation had come to the mute young man like the paraclete, bearing the gift of tongues.

Here in the bar, it grew plainer and plainer that the colonel did not regard himself as an anti-Semite but merely as a heavy thinker. The idea that I considered him anti-Semitic sincerely outraged his feelings. "Prejudice" was the last trait he could have imputed to himself. He looked on me, almost respectfully, as a "Jew-lover," a kind of being he had heard of but never actually encountered, like a centaur or a Siamese twin, and the interest of relating this prodigy to the natural state of mankind overrode any personal distaste. There I sat, the exception which was "proving" or testing the rule, and he kept pressing me for details of my history that might explain my deviation in terms of the norm. On my side, of course, I had become fiercely resolved that he would learn nothing from me that would make it possible for him to dismiss my anti-anti-Semitism as the product of special circumstances: I was stubbornly sitting on the fact of my Jewish grandmother like a hen on a golden egg. I was bent on making *him* see himself as a monster, a deviation, a heretic from Church and State. Unfortunately, the colonel, owing perhaps to his military training, had not the glimmering of an idea of what democracy meant; to him, it was simply a slogan that was sometimes useful in war. The notion of an ordained inequality was to him "scientific."

"Honestly," he was saying in lowered tones, as our drinks were taken away and the waitress set down my sandwich and his corned-beef hash, "don't you, brought up the way you were, feel about them the way I do? Just between ourselves, isn't there a sort of inborn feeling of horror that the very word, Jew, suggests?" I shook my head, roundly. The idea of an *innate* anti-Semitism was in keeping with the rest of the colonel's thought, yet it shocked me more than anything he had yet said. "No," I sharply replied. "It doesn't evoke any feeling one way or the other." "Honest Injun?" said the colonel. "Think back: when you were a kid, didn't the word, Jew, make you feel sick?" There was a dreadful sincerity about this that made me answer in an almost kindly tone. "No, truthfully, I assure you. When we were children, we learned to call the old-clothes man a sheeny, but that was just a dirty word to us, like 'Hun' that we used to call after workmen we thought were Germans."

"I don't get it," pondered the colonel, eating a pickle. "There must be something wrong with you. Everybody is born with that feeling. It's natural; it's part of nature." "On the contrary," I said. "It's something very unnatural that you must have been taught as a child." "It's not something you're *taught*," he protested. "You must have been," I said. "You simply don't remember it. In any case, you're a man now; you must rid yourself of that feeling. It's psychopathic, like that horrible young man on the train." "You thought he was crazy?" mused the colonel, in an idle, dreamy tone. I shrugged my shoulders. "Of course. Think of his color. He was probably just out of a mental institution. People don't get that tattletale gray except in prison or mental hospitals." The colonel suddenly grinned. "You might be right," he said. "He was quite a case." He chuckled.

I leaned forward. "You know, Colonel," I said quickly, "anti-Semitism is contrary to the Church's teaching. God will make you do penance for hating the Jews. Ask your priest; he'll tell you I'm right. You'll have a long spell in Purgatory, if you don't rid yourself of this sin. It's a deliberate violation of Christ's commandment, 'Love thy neighbor.' The Church holds that the Jews have a sacred place in God's design. Mary was a Jew and Christ was a Jew. The Jews are under God's special protection. The Church teaches that the millennium can't come until the conversion of the Jews; therefore, the Jews must be preserved that the Divine Will may be accomplished. Woe to them that harm them, for they controvert God's Will!" In the course of speaking, I had swept myself away with the solemnity of the doctrine. The Great Reconciliation between God and His chosen people, as envisioned by the Evangelist, had for me at that moment a piercing, majestic beauty, like some awesome Tintoretto. I saw a noble spectacle of blue sky, thronged with gray clouds, and a vast white desert, across which God and Israel advanced to meet each other, while below in hell the demons of disunion shrieked and gnashed their teeth.

"Hell," said the colonel, jovially, "I don't believe in all that. I lost my faith when I was a kid. I saw that all this God stuff was a lot of bushwa." I gazed at him in stupefaction. His confidence had completely returned. The blue eyes glittered debonairly, the eagles glittered; the narrow polished head cocked and listened to itself like a trilling bird. I was up against an airman with a bird's-eye view, a man who believed in nothing but the law of kind: the epitome of godless materialism. "You still don't hold with that bunk?" the colonel inquired in an undertone, with an expression of stealthy curiosity. "No," I confessed, sad to admit to a meeting of minds. "You know what got me?" exclaimed the colonel. "That birth-control stuff. Didn't it kill you?" I made a neutral sound. "I was beginning to play around," said the colonel, with a significant beam of the eye, "and I just couldn't take that guff. When I saw through the birth-control talk, I saw through the whole thing. They claimed it was against nature, but I claim, if that's so, an operation's against nature. I told my old man that when he was having his kidney stones out. You ought to have heard him yell!" A rich, reminiscent satisfaction dwelt in the colonel's face.

This period of his life, in which he had thrown off the claims of the spiritual and adopted a practical approach, was evidently one of those "turning points" to which a man looks back with pride. He lingered over the story of his break with church and parents with a curious sort of heat, as though the flames of old sexual conquests stirred within his body at the memory of those old quarrels. The looks he rested on me, as a sharer of that experience, grew more and more lickerish and assaying. "What got *you* down?" he finally inquired, settling back in his chair and pushing his coffee cup aside. "Oh," I said wearily, "it's a long story. You can read it when it's published." "You're an author?" cried the colonel, who was really very slow-witted. I nodded, and the colonel regarded me afresh. "What do you write? Love stories?" He gave a half-wink. "No," I said. "Various things. Articles. Books. Highbrowish stories." A suspicion darkened in the colonel's sharp face. "That McCarthy," he said. "Is that your pen name?" "Yes," I said, "but it's my real name too. It's the name I write under *and* my maiden name." The colonel digested this thought. "Oh," he concluded.

A new idea seemed to visit him. Quite cruelly, I watched it take possession. He was thinking of the power of the press and the indiscretions of other military figures, who had been rewarded with demotion. The consciousness of the uniform he wore appeared to seep uneasily into his body. He straightened his shoulders and called thoughtfully for the check. We paid in silence, the colonel making no effort to forestall my dive into my pocketbook. I should not have let him pay in any case, but it startled me that he did not try to do so, if only for reasons of vanity. The whole business of paying, apparently, was painful to him; I watched his facial muscles contract as he pocketed the change and slipped two dimes for the waitress onto the table, not daring quite to hide them under the coffee cup—he had short-changed me on the bill and the tip, and we both knew it. We walked out into the steaming station and I took my baggage out of the checking locker. The colonel carried my suitcase and we strolled along without speaking. Again, I felt horribly embarrassed for him. He was meditative, and I supposed that he too was mortified by his meanness about the tip.

"Don't get me wrong," he said suddenly, setting the suitcase down and turning squarely to face me, as though he had taken a big decision. "I may have said a few things back there about the Jews getting what they deserved in Germany." I looked at him in surprise; actually, he had not said that to me. Perhaps he had let it drop in the club car after I had left. "But that doesn't mean I approve of Hitler." "I should hope not," I said. "What I mean is," said the colonel, "that they probably gave the Germans a lot of provocation, but that doesn't excuse what Hitler did." "No," I said, somewhat ironically, but the colonel was unaware of anything satiric in the air. His face was grave and determined; he was sorting out his philosophy for the record. "I mean, I don't approve of his methods," he finally stated. "No," I agreed. "You mean, you don't approve of the gas chamber." The colonel shook his head very severely. "Absolutely not! That was terrible." He shuddered and drew out a handkerchief and slowly wiped his brow. "For God's sake," he said, "don't get me wrong. I think they're human beings." "Yes," I assented, and we walked along to my track. The colonel's spirits lifted, as though, having stated his credo, he had both got himself in line with public policy and achieved an autonomous thought. "I mean," he resumed, "you may not care for them, but that's not the same as killing them, in cold blood, like that." "No, Colonel," I said.

He swung my bag onto the car's platform and I climbed up behind it. He stood below, smiling, with upturned face. "I'll look for your article," he cried, as the train whistle blew. I nodded, and the colonel waved, and I could not stop myself from waving back at him and even giving him the corner of a smile. After all, I said to myself, looking down at him, the colonel was "a human being." There followed one of those inane intervals in which one prays for the train to leave. We both glanced at our watches. "See you some time," he called. "What's your married name?" "Broadwater," I called back. The whistle blew again. "Broadwater?" shouted the colonel, with a dazed look of unbelief and growing enlightenment; he was not the first person to hear it as a Jewish name, on the model of Goldwater. "B-r-o-a-d," I began, automatically, but then I stopped. I disdained to spell it out for him; the victory was his. "One of the chosen, eh?" his brief grimace seemed to commiserate. For the last time, and in the final fullness of understanding, the hawk eye patrolled the green dress, the duster, and the earrings; the narrow flue of his nostril contracted as he curtly turned away. The train commenced to move.

11

Settling the Colonel's Hash

Mary McCarthy

February, 1954

Seven years ago, when I taught in a progressive college, I had a pretty girl student in one of my classes who wanted to be a short-story writer. She was not studying writing with me, but she knew that I sometimes wrote short stories, and one day, breathless and glowing, she came up to me in the hall, to tell me that she had just written a story that her writing teacher, a Mr. Converse, was terribly excited about. "He thinks it's wonderful," she said, "and he's going to help me fix it up for publication."

I asked what the story was about; the girl was a rather simple being who loved clothes and dates. Her answer had a deprecating tone. It was just about a girl (herself) and some sailors she had met on the train. But then her face, which had looked perturbed for a moment, gladdened.

"Mr. Converse is going over it with me and we're going to put in the symbols."

Another girl in the same college, when asked by us in her sophomore orals why she read novels (one of the pseudoprofound questions that ought never to be put) answered in a defensive flurry: "Well, *of course,* I don't read them to find out what happens to the hero."

At the time, I thought these notions were peculiar to progressive education: it was old-fashioned or regressive to read a novel to find out what happens to the hero or to have a mere experience empty of symbolic patterns. But now I discover that this attitude is quite general, and that readers and students all over the country are in a state of apprehension, lest they read a book or story literally and miss the presence of a symbol. And like everything in America, this search for meanings has become a socially competitive enterprise; the best reader is the one who detects the most symbols in a given stretch of prose. And the benighted reader who fails to find any symbols humbly assents when they are pointed out to him; he accepts his mortification.

I had no idea how far this process had gone until last spring, when I began to get responses to a story I had published in *Harper's*. I say "story" because that was what it was called by *Harper's*. I myself would not know quite what to call it; it was a piece of reporting or a fragment of autobiography—an account of my meeting with an anti-Semitic army colonel. It began in the club car of a train going to St. Louis; I was wearing an apple-green shirtwaist and a dark-green skirt and pink earrings; we got into an argument about the Jews. The colonel was a rather dapper, flashy kind of Irish-American with a worldly blue eye; he took me, he said, for a sculptress, which made me feel, to my horror, that I looked Bohemian and therefore rather suspect. He was full of the usual profound cliches that anti-Semites air, like original epigrams, about the Jews; that he could tell a Jew, that they were different from other people, that you couldn't trust them in business, that some of his best friends were Jews, that he distinguished between a Jew and a kike, and finally that, of course, he didn't agree with Hitler: Hitler went too far; the Jews were human beings.

All the time we talked, and I defended the Jews, he was trying to get my angle, as he called it; he thought it was abnormal for anybody who wasn't Jewish not to feel as he did. As a matter of fact, I have a Jewish grandmother, but I decided to keep this news to myself: I did not want the colonel to think that I had any interested reason for speaking on behalf of the Jews, that is, that I was prejudiced. In the end, though, I got my comeuppance. Just as we were parting, the colonel asked me my married name, which is Broadwater, and the whole mystery was cleared up for him, instantly; he supposed I was married to a Jew and that the name was spelled B-r-o-d-w-a-t-e-r. I did not try to enlighten him; I let him think what he wanted; in a certain sense, he was right; he had unearthed my Jewish grandmother or her equivalent. There were a few details that I must mention to make the next part clear: in my car, there were two nuns, whom I talked to as a distraction from the colonel and the moral problems he raised. He and I finally had lunch together in the St. Louis railroad station, where we continued the discussion. It was a very hot day. I had a sandwich; he had roast-beef hash. We both had an old-fashioned.

The whole point of this "story" was that it really happened; it is written in the first person; I speak of myself in my own name, McCarthy; at the end, I mention my husband's name, Broadwater. When I was thinking about writing the story, I decided not to treat it fictionally; the chief interest, I felt, lay in the fact that it happened, in real life, last summer, to the writer herself, who was a good deal at fault in the incident. I wanted to embarrass myself and, if possible, the reader too.

Yet, strangely enough, many of my readers preferred to think of this account as fiction. I still meet people who ask me, confidentially, "That story of yours about the colonel—was it really true?" It seemed to them perfectly natural that I would write a fabrication, in which I figured under my own name, and sign it, though in my eyes this would be like perjuring yourself in court or forging checks. Shortly after the "story" was published, I got a kindly letter from a man in Mexico, in which he criticized the menu from an artistic point of view: he thought salads would be better for hot weather and it would be more in character for the narrator-heroine to have a Martini. I did not answer the letter, though I was moved to, because I had the sense that he would not understand the distinction between what *ought* to have happened and what *did* happen.

Then in April I got another letter, from an English teacher in a small college in the Middle West, that reduced me to despair. I am going to cite it at length.

"My students in freshman English chose to analyze your story, 'Artists in Uniform,' from the March issue of *Harper's*. For a week

I heard oral discussions on it and then the students wrote critical analyses. In so far as it is possible, I stayed out of their discussions, encouraging them to read the story closely with your intentions as a guide to their understanding. Although some of them insisted that the story has no other level than the realistic one, most of them decided it has symbolic overtones.

"The question is: how closely do you want the symbols labeled? They wrestled with the nuns, the author's two shades of green with pink accents, with the 'materialistic godlessness' of the colonel. . . . A surprising number wanted exact symbols; for example, they searched for the significance of the colonel's eating hash and the author eating a sandwich. . . . From my standpoint, the story was an entirely satisfactory springboard for understanding the various shades of prejudice, for seeing how much of the artist goes into his painting. If it is any satisfaction to you, our campus was alive with discussions about 'Artists in Uniform.' We liked the story and we thought it amazing that an author could succeed in making readers dislike the author—for a purpose, of course!"

I probably should have answered this letter, but I did not. The gulf seemed to me too wide. I could not applaud the backward students who insisted that the story had no other level than the realistic one without giving offense to their teacher, who was evidently a well-meaning person. But I shall try now to address a reply, not to this teacher and her unfortunate class, but to a whole school of misunderstanding. There were no symbols in this story; there was no deeper level. The nuns were in the story because they were on the train; the contrasting greens were the dress I happened to be wearing; the colonel had hash because he had hash; materialistic godlessness meant just what it means when a priest thunders it from the pulpit—the phrase, for the first time, had meaning for me as I watched and listened to the colonel.

But to clarify the misunderstanding, one must go a little further and try to see what a literary symbol is. Now in one sense, the colonel's hash and my sandwich can be regarded as symbols; that is, they typify the colonel's food tastes and mine. (The man in Mexico had different food tastes which he wished to interpose into our reality.) The hash and the sandwich might even be said to show something very obvious about our characters and bringing-up, or about our sexes; I was a woman, he was a man. And though on another day I might have ordered hash myself, that day I did not, because the colonel and I, in our disagreement, were polarizing each other.

The hash and the sandwich, then, could be regarded as symbols of our disagreement, almost conscious symbols. And underneath our discussion of the Jews, there was a thin sexual current running, as there always is in such random encounters or pickups (for they have a strong suggestion of the illicit). The fact that I ordered something conventionally feminine and he ordered something conventionally masculine represented, no doubt, our awareness of a sexual possibility; even though I was not attracted to the colonel, nor he to me, the circumstances of our meeting made us define ourselves as a woman and man.

The sandwich and the hash were our provisional, ad hoc symbols of ourselves. But in this sense all human actions are symbolic because they represent the person who does them. If the colonel had ordered a fruit salad with whipped cream, this too would have represented him in some way; given his other traits, it would have pointed to a complexity in his character that the hash did not suggest.

In the same way, the contrasting greens of my dress were a symbol of my taste in clothes and hence representative of me—all too representative, I suddenly saw, in the club car, when I got an "artistic" image of myself flashed back at me from the men's eyes. I had no wish to stylize myself as an artist, that is, to parade about as a symbol of flamboyant unconventionality, but apparently I had done so unwittingly when I picked those colors off a rack, under the impression that they suited me or "expressed my personality" as salesladies say.

My dress, then, was a symbol of the perplexity I found myself in with the colonel; I did not want to be categorized as a member of a peculiar minority—an artist or a Jew; but brute fate and the colonel kept resolutely cramming me into both those uncomfortable pigeonholes. I wished to be regarded as ordinary or rather as universal, to be anybody and therefore everybody (that is, in one sense, I wanted to be on the colonel's side, majestically above minorities); but every time the colonel looked at my dress and me in it with my pink earrings I shrank to minority status, and felt the dress in the heat shriveling me, like the shirt of Nessus, the centaur, that consumed Hercules.

But this is not what the students meant when they wanted the symbols "labeled." They were searching for a more recondite significance than that afforded by the trite symbolism of ordinary life, in which a dress is a social badge. They supposed that I was engaging in literary or artificial symbolism which would lead the reader out of the confines of reality into the vast fairy tale of myth, in which the color green would have an emblematic meaning (or did the two greens signify for them what the teacher calls "shades" of prejudice), and the colonel's hash, I imagine, would be some sort of Eucharistic mincemeat.

Apparently, the presence of the nuns assured them that there were overtones of theology; it did not occur to them (a) that the nuns were there because pairs of nuns are a standardized feature of summer Pullman travel, like crying babies, and perspiring businessmen in the club car, and (b) that if I thought the nuns worth mentioning, it was also because of something very simple and directly relevant: the nuns and the colonel and I all had something in common—we had all at one time been Catholics—and I was seeking common ground with the colonel, from which to turn and attack his position.

In any account of reality, even a televised one, which comes closest to being a literal transcript or replay, some details are left out as irrelevant (though nothing is really irrelevant). The details that are not eliminated have to stand as symbols of the whole, like stenographic signs, and of course there is an art of selection, even in a newspaper account: the writer, if he has any ability, is looking for the revealing detail that will sum up the picture for the reader in a flash of recognition.

But the art of abridgment and condensation, which is familiar to anybody who tries to relate an anecdote, or give a direction—the art of natural symbolism, which is at the basis of speech and all representation—has at bottom a centripetal intention. It hovers over an object, an event, or series of events and tries to declare what it is. Analogy (that is, comparison to other objects) is inevitably one of its methods. "The weather was soupy," i.e., like soup.

"He wedged his way in," i.e., he had to enter, thin edge first, as a wedge enters, and so on. All this is obvious. But these metaphorical aids to communication are a far cry from literary symbolism, as taught in the schools and practiced by certain fashionable writers. Literary symbolism is centrifugal and flees from the object, the event, into the incorporeal distance, where concepts are taken for substance and floating ideas and archetypes assume a hieratic authority.

In this dream-forest, symbols become arbitrary; all counters are interchangeable; anything can stand for anything else. The colonel's hash can be a Eucharist or a cannibal feast or the banquet of Atreus, or all three, so long as the actual dish set before the actual man is disparaged. What is depressing about this insistent symbolization is the fact that while it claims to lead to the infinite, it quickly reaches very finite limits—there are only so many myths on record, and once you have got through Bulfinch, the Scandinavian, and the Indian, there is not much left. And if all stories reduce themselves to myth and symbol, qualitative differences vanish, and there is only a single, monotonous story.

American fiction of the symbolist school demonstrates this mournful truth, without precisely intending to. A few years ago, when the mode was at its height, chic novels and stories fell into three classes: those which had a Greek myth for their framework, which the reader was supposed to detect, like finding the faces in the clouds in old newspaper puzzle contests; those which had symbolic modern figures, dwarfs, hermaphrodites, and cripples, illustrating maiming and loneliness; and those which contained symbolic animals, cougars, wild cats, and monkeys. One young novelist, a product of the Princeton school of symbolism, had all three elements going at once, like the ringmaster of a three-ring circus, with the freaks, the animals, and the statues.

The quest for symbolic referents had as its object, of course, the deepening of the writer's subject and the reader's awareness. But the result was paradoxical. At the very moment when American writing was penetrated by the symbolic urge, it ceased to be able to create symbols of its own. Babbitt, I suppose, was the last important symbol to be created by an American writer; he gave his name to a type that henceforth would be recognizable to everybody. He passed into the language. The same thing could be said, perhaps, though to a lesser degree, of Caldwell's Tobacco Road, Eliot's Prufrock, and possibly of Faulkner's Snopeses. The discovery of new symbols is not the only function of a writer, but the writer who cares about this must be fascinated by reality itself, as a butterfly collector is fascinated by the glimpse of a new specimen. Such a specimen was Mme. Bovary or M. Homais or M. de Charlus or Jupien; these specimens were precious to their discoverers, not because they repeated an age-old pattern but because their markings were new. Once the specimen has been described, the public instantly spots other examples of the kind, and the world seems suddenly full of Babbitts and Charlus, where none had been noted before.

A different matter was Joyce's Mr. Bloom. Mr. Bloom can be called a symbol of eternal recurrence—the wandering Jew, Ulysses the voyager—but he is a symbol thickly incarnate, fleshed out in a Dublin advertising canvasser. He is not *like* Ulysses or vaguely suggestive of Ulysses; he is Ulysses, circa 1905. Joyce evidently believed in a cyclical theory of history, in which everything repeated itself; he also subscribed in youth to the doctrine that declares that the Host, a piece of bread, is also God's body and blood. How it can be both things at the same time, transubstantially, is a mystery, and Mr. Bloom is just such a mystery: Ulysses in the visible appearance of a Dublin advertising canvasser.

Mr. Bloom is not a symbol of Ulysses, but Ulysses-Bloom together, one and indivisible, symbolize or rather demonstrate eternal recurrence. I hope I make myself clear. The point is transubstantiation: Bloom and Ulysses are transfused into each other and neither reality is diminished. Both realities are locked together, like the protons and neutrons of an atom. *Finnegans Wake* is a still more ambitious attempt to create a fusion, this time a myriad fusion, and to exemplify the mystery of how a thing can be of itself and at the same time be something else. The word is many and it is also one.

But the clarity and tension of Joyce's thought brought him closer in a way to the strictness of allegory than to the diffuse practices of latter-day symbolists. In Joyce, the equivalences and analogies are very sharp and distinct, as in a pun, and the real world is almost querulously audible, like the voices of the washerwomen on the Liffey that come into Earwicker's dream. But this is not true of

Joyce's imitators or of the imitators of his imitators, for whom reality is only a shadowy pretext for the introduction of a whole *corps de ballet* of dancing symbols in mythic draperies and animal skins.

Let me make a distinction. There are some great writers, like Joyce or Melville, who have consciously introduced symbolic elements into their work; and there are great writers who have written fables or allegories. In both cases, the writer makes it quite clear to the reader how he is to be read; only an idiot would take *Pilgrim's Progress,* for a realistic story, and even a young boy, reading *Moby Dick,* realizes that there is something more than whale-fishing here, though he may not be able to name what it is. But the great body of fiction contains only what I have called *natural symbolism,* in which selected events represent or typify a problem, a kind of society or psychology, a philosophical theory, in the same way that they do in real life. What happens to the hero becomes of the highest importance. This symbolism needs no abstruse interpretation, and abstruse interpretation will only lead the reader away from the reality that the writer is trying to press on his attention.

I shall give an example or two of what I mean by natural symbolism and I shall begin with a rather florid one: Henry James' *The Golden Bowl.* This is the story of a rich American girl who collects European objects. One of these objects is a husband, Prince Amerigo, who proves to be unfaithful. Early in the story, there is a visit to an antique shop in which the Prince picks out a gold bowl for his fiancée and finds, to his annoyance, that it is cracked. It is not hard to see that the cracked bowl is a symbol, both of the Prince himself, who is a valuable antique but a little flawed, morally, and also of the marriage, which represents an act of acquisition or purchase on the part of the heroine and her father. If the reader should fail to notice the analogy, James calls his attention to it in the title.

I myself would not regard this symbol as necessary to this particular history; it seems to me, rather, an ornament of the kind that was fashionable in the architecture and interior decoration of the period, like stylized sheaves of corn or palms on the façade of a house. Nevertheless, it is handsome and has an obvious appropriateness to the theme. It introduces the reader into the Gilded Age attitudes of the novel. I think there is also a scriptural echo in the title that conveys the idea of punishment. But having seen and felt the weight of meaning that James put into this symbol, one must not be tempted to press further and look at the bowl as a female sex symbol, a chalice, a Holy Grail, and so on; a book is not a pious excuse for reciting a litany of associations.

My second example is from Tolstoy's *Anna Karenina.* Toward the beginning of the novel, Anna meets the man who will be her lover, Vronsky, on the Moscow–St. Petersburg express; as they meet, there has been an accident; a workman has been killed by the train. This is the beginning of Anna's doom, which is completed when she throws herself under a train and is killed; and the last we see of Vronsky is in a train, with a toothache; he is off to the wars. The train is necessary to the plot of the novel, and I believe it is also symbolic, both of the iron forces of material progress that Tolstoy hates so and that played a part in Anna's moral destruction, and also of those iron laws of necessity and consequence that govern human action when it remains on the sensual level.

One can read the whole novel, however, without being conscious that the train is a symbol; we do not have to "interpret" or feel the import of doom and loneliness in the train's whistle—the same import we ourselves can feel when we hear a train whistle blow in the country, even today. Tolstoy was a deeper artist than James, and we cannot be sure that the train was a conscious device with him. The appropriateness to Anna's history may have been only a *felt* appropriateness; everything in Tolstoy has such a supreme naturalness that one shrinks from attributing contrivance to him, as if it were a sort of fraud. Yet he worked very hard on his novels—I forget how many times Countess Tolstoy copied out *War and Peace* by hand.

The impression one gets from his diaries is that he wrote by ear; he speaks repeatedly, even as an old man, of having to start a story over again because he has the wrong tone, and I suspect that he did not think of the train as a symbol but that it sounded "right" to him, because it was, in that day, an almost fearsome emblem of ruthless and impersonal force, not only to a writer of genius but to the poorest peasant in the fields. And in Tolstoy's case I think it would be impossible, even for the most fanciful critic, to extricate the train from the novel and try to make it say something that the novel does not say directly. Every detail in Tolstoy has an almost cruel and viselike meaningfulness and truth to itself that make it tautological to talk of symbolism; he was a moralist and to him the tiniest action, even the curiosities of physical appearance, Vronsky's bald spot, the small white hands of Prince Andrei, told a moral tale.

It is now considered very old-fashioned and tasteless to speak of an author's "philosophy of life" as something that can be harvested from his work. Actually, most of the great authors did have a "philosophy of life" which they were eager to communicate to the public; this was one of their motives for writing. And to disentangle a moral philosophy from a work that evidently contains one is far less damaging to the author's purpose and the integrity of his art than to violate his imagery by symbol-hunting, as though reading a novel were a sort of paperchase.

The images of a novel or a story belong, as it were, to a family, very closely knit and inseparable from each other; the parent "idea" of a story or a novel generates events and images all bearing a strong family resemblance. And to understand a story or a novel, you must look for the parent "idea," which is usually in plain view, if you read quite carefully and literally what the author says.

I will go back, for a moment, to my own story, to show how this can be done. Clearly, it is about the Jewish question, for that is what the people are talking about. It also seems to be about artists, since the title is "Artists in Uniform." Then there must be some relation between artists and Jews. What is it? They are both minorities that other people claim to be able to recognize by their appearance. But artists and Jews do not care for this categorization; they want to be universal, that is, like everybody else. They do not want to wear their destiny as a badge, as the soldier wears his uniform. But this aim is really hopeless, for life has formed them as Jews or artists, in a way that immediately betrays them to the majority they are trying to melt into. In my conversation with the colonel, I was endeavoring to play a double game. I was trying to force him into a minority by treating anti-Semitism as an aberration, which, in fact, I believe it is. On his side, the colonel resisted this attempt and tried to show that anti-Semitism was normal, and he was normal, while I was the queer one. He declined to be categorized as anti-Semite; he regarded himself as an independent thinker, who by a happy chance thought the same as everybody else.

I imagined I had a card up my sleeve; I had guessed that the colonel was Irish (i.e., that he belonged to a minority) and presumed that he was a Catholic. I did not see how he could possibly guess that I, with my Irish name and Irish appearance, had a Jewish grandmother in the background. Therefore when I found I had not convinced him by reasoning, I played my last card; I told him that the Church, his Church, forbade anti-Semitism. I went even further; I implied that God forbade it, though I had no right to do this, since I did not believe in God, but was only using Him as a whip to crack over the colonel, to make him feel humble and inferior, a raw Irish Catholic lad under discipline. But the colonel, it turned out, did not believe in God, either, and I lost. And since, in a sense, I had been cheating all along in this game we were playing, I had to concede the colonel a sort of moral victory in the end; I let him think that my husband was Jewish and that that "explained" everything satisfactorily.

Now there are a number of morals or meanings in this little tale, starting with the simple one: don't talk to strangers on a train. The chief moral or meaning (what I learned, in other words, from this experience) was this: you cannot be a universal unless you accept the fact that you are a singular, that is, a Jew or an artist or what-have-you. What the colonel and I were discussing, and at the same time illustrating and enacting, was the definition of a human being. I was trying to be something better than a human being; I was trying to be the voice of pure reason; and pride went before a fall. The colonel, without trying, was being something worse than a human being, and somehow we found ourselves on the same plane—facing each other, like mutually repellent twins. Or, put in another way: it is dangerous to be drawn into discussions of the Jews with anti-Semites: you delude yourself that you are spreading light, but you are really sinking into muck; if you endeavor to be dispassionate, you are really claiming for yourself a privileged position, a little mountain top, from which you look down, impartially, on both the Jews and the colonel.

Anti-Semitism is a horrible disease from which nobody is immune, and it has a kind of evil fascination that makes an enlightened person draw near the source of infection, supposedly in a scientific spirit, but really to sniff the vapors and dally with the possibility. The enlightened person who lunches with the colonel in order, as she tells herself, to improve him, is cheating herself, having her cake and eating it. This attempted cheat, on my part, was related to the question of the artist and the green dress; I wanted to be an artist but not to pay the price of looking like one, just as I was willing to have Jewish blood but not willing to show it, where it would cost me something—the loss of superiority in an argument.

These meanings are all there, quite patent, to anyone who consents to look *into* the story. They were *in* the experience itself, waiting to be found and considered. I did not perceive them all at the time the experience was happening; otherwise, it would not have taken place, in all probability—I should have given the colonel a wide berth. But when I went back over the experience, in order to write it, I came upon these meanings, protruding at me, as it were, from the details of the occasion. I put in the green dress and my mortification over it because they were part of the truth, just as it had occurred, but I did not see how they were related to the general question of anti-Semitism and my grandmother until they *showed* me their relation in the course of writing.

Every short story, at least for me, is a little act of discovery. A cluster of details presents itself to my scrutiny, like a mystery that I will understand in the course of writing or sometimes not fully until afterward, when, if I have been honest and listened to these details carefully, I will find that they are connected and that there is a coherent pattern. This pattern is *in* experience itself; you do not impose it from the outside and if you try to, you will find that the story is taking the wrong tact, dribbling away from you into artificiality or inconsequence. A story that you do not learn something from while you are writing it, that does not illuminate something for you, is dead, finished before you started it. The "idea" of a story is implicit in it, on the one hand; on the other, it is always ahead of the writer, like a form dimly discerned in the distance; he is working *toward* the "idea."

It can sometimes happen that you begin a story thinking that you know the "idea" of it and find, when you are finished, that you have said something quite different and utterly unexpected to you. Most writers have been haunted all their lives by the "idea" of a story or a novel that they think they want to write and see very clearly: Tolstoy always wanted to write a novel about the Decembrists and instead, almost against his will, wrote *War and Peace;* Henry James thought he wanted to write a novel about Napoleon. Probably these ideas for novels were too set in their creators' minds to inspire creative discovery.

In any work that is truly creative, I believe, the writer cannot be omniscient in advance about the effects that he proposes to produce. The suspense in a novel is not only in the reader, but in the novelist himself, who is intensely curious too about what will happen to the hero. Jane Austen may know in a general way that Emma will marry Mr. Knightley in the end (the reader knows this too, as a matter of fact); the suspense for the author lies in the how, in the twists and turns of circumstance, waiting but as yet unknown, that will bring the consummation about. Hence, I would say to the student of writing that outlines, patterns, arrangements of symbols may have a certain usefulness at the outset for some kind of minds, but in the end they will have to be scrapped. If the story does not contradict the outline, overrun the pattern, break the symbols, like an insurrection against authority, it is surely a still birth. The natural symbolism of reality has more messages to communicate than the dry Morse code of the disengaged mind.

The tree of life, said Hegel, is greener than the tree of thought; I have quoted this before but I cannot forbear from citing it again in this context. This is not an incitement to mindlessness or an endorsement of realism in the short story (there are several kinds of reality, including interior reality); it means only that the writer must be, first of all, a listener and observer, who can pay attention to reality, like an obedient pupil, and who is willing, always, to be surprised by the messages reality is sending through to him. And if he gets the messages correctly he will not have to go back and put in the symbols; he will find that the symbols are there, staring at him significantly from the commonplace.

12

Unsettling the Colonel's
Hash "Fact" in Autobiography

Darrel Mansell

"Artists in Uniform: A Story by Mary McCarthy" appeared in
Harper's Magazine (March 1953). The piece describes an encounter in the club car of a train, and later in a station restaurant,
between a woman and an anti-Semitic air force colonel. In 1954
McCarthy published an essay, "Settling the Colonel's Hash."[1] She
discusses various questions, comments, and interpretations of "Artists in Uniform" passed on to her by readers, and goes on to give
her own "idea" of the work, its "chief moral or meaning" to her.
Here is an author's sincere and genial attempt to set her readers
right on what she meant. Such an attempt always raises fundamental issues concerning the nature of art, and below I want to explore
one of them.

According to McCarthy, her readers have been misled from the
beginning. For it was the editors of the magazine who were responsible for the subtitle that refers to "Artists in Uniform" as a
"story." "I myself," she says, "would not know quite what to call it;
it was a piece of reporting or a fragment of autobiography." Indeed, the

> whole point of this "story" was that it really happened; it is written
> in the first person; I speak of myself in my own name, Mc-
> Carthy. . . . When I was thinking about writing the story, I decided

not to treat it fictionally; the chief interest, I felt, lay in the fact that
it happened, in real life, . . . to the writer herself.

The fact that the events "happened, in real life" makes all the
difference to her. For when her readers asked whether details like
the colonel's having hash for lunch were "symbolic" of something
in the same way such details might be symbolic in fiction ("a Eucharist or a cannibal feast or the banquet of Atreus"), she felt the
correct response was to say that there "were no symbols in this
story; there was no deeper level. The nuns were in the story because they were on the train; . . . the colonel had hash because he
had hash."

McCarthy is making the assumption we all make except in our
stern philosophical moments: that there are two more or less separate literary kinds, "autobiography" and "narrative fiction"; and
that these are separate because the one deals substantially with
fact, what "happened, in real life," no matter how "artistic" the
presentation, and the other deals substantially with what did not
happen but was only "imagined," no matter how convincingly
"real." Our "autobiography" and "fiction" categories thus seem
themselves to be manifestations of an even more fundamental distinction, between declarations of "fact" and some other broad category of fictitious or otherwise imaginative works.

The distinction appears in Coleridge's *Biographia Literaria:*
there is a "species of composition, which is opposed to works of
science, by proposing for its *immediate* object pleasure, not truth";[2]
and this general idea is absolutely crucial in modern literary society.
There is a Great Divide that runs down the center of the literary
map. Works like autobiography which deal, for the most part, with
what we can suppose "happened, in real life"—these lie somewhere high on one slope that terminates in the smooth waters of
pure scientific discourse. The various forms of "fiction" lie somewhere on the other slope, at the bottom of which is whatever one
could conceive to be purely imaginative literature, "the foam / Of
perilous seas, in faery lands forlorn." Thus René Wellek and Robert Penn Warren distinguish between the "scientific uses of language" and the "literary" uses; Cleanth Brooks distinguishes between the "terms of science" and poetic language; Northrop Frye
between "descriptive" writing and literature; I. A. Richards be-

tween the "undistorted references" of science and *"fictions"*; and Ralph W. Rader refers typically to that "fundamental contrast between the fictional and the factual narrative modes."[3]

This distinction helps a librarian decide where to put *The Education of Henry Adams* and *David Copperfield*. The one book may not entirely conform to what we know of Adams's life, and the other may have drawn much of its material from Dickens's own. Still, one falls under "fact" and narrates events we suppose "happened" to some controlling degree; the other is on the opposite side of the Great Divide and narrates events we suppose are mostly imagined.

Granted, there are borderline cases. Is *Sartor Resartus* fiction, Carlyle's veiled autobiography, or perhaps something else?[4] But we can, and do, solve such problems by turning them into a simple matter of proportion. We tote up the specific details that seem fictional (the hero's name, Teufelsdröckh) and the ones that seem factual (Carlyle said elsewhere that his character's spiritual conversion resembled his own), and decide to our own satisfaction which predominates. We still end up making a pretty and uncomplicated distinction. As Frye puts it, "an autobiography coming into a library would be classified as nonfiction if the librarian believed the author, and as fiction if she thought he was lying."

Thus that fundamental critical distinction between statements of truth and fiction has always put autobiography on one side of the Divide, and almost everything we think of as "literature" on the other. Autobiography becomes a species of scientific discourse— and indeed the Library of Congress classification system makes Autobiography a subclass of Biography, and Biography a subclass of the "Auxiliary Sciences of History."[5] Likewise, Wayne Shumaker makes autobiography, along with biography and history, one of the "factual literary types"; Barrett J. Mandel lists "avowed truth" as one of the features that set autobiography off from other kinds of literature; and Käte Hamburger makes the general statement that "an innate characteristic of every first-person narrative" is that "it posits itself as non-fiction, i.e., as a historical document," a "reality statement."[6] So, we think of autobiography as a more or less direct relation of factual truth; and we even pass judgment on autobiographies according to whether they stick to the facts, the way things "happened, in real life." Roy Pascal, for instance, points out that when we read an autobiography, "We like to ask, does the author's representation of himself . . . correspond to what we can get to know of him through other evidence? It is a question that can never be asked regarding a work of art."

Autobiography therefore is not "art," not "literature," not "fiction." It is on the side of biography, history, and the truth-telling sciences. Its truth is occasionally conceived as somehow "higher" than a mere conscientious relation of fact;[7] but there is absolutely no way to free autobiography significantly from its obligation to fact. Here we are at the core of what seems to differentiate this genre from "fiction." We really do believe that autobiography is somehow obligated to fact, what happened in real life, in a way that "literature" is not. We believe that fact is fact, and fiction is something else. We have not progressed far beyond that simple, inexpugnable distinction Sir Philip Sidney took for granted long ago: "think I none so simple would say that *Esope* lyed in the tales of his beasts: for who thinks that *Esope* writ it for actually true were well worthy to haue his name cronicled among the beastes hee writeth of."[8]

But what happens when we examine this distinction, not as a belief, but as a critical, or aesthetic, or even "philosophical" principle for separating autobiography from fiction? The principle erodes, until there is nothing left.

To begin with, how do we decide whether to take something as fact or fiction? Suppose somebody writes on a piece of paper, "I was born on a Friday, at twelve o'clock at night." Fact or fiction? There is of course no way of telling (in this case it is Dickens's *David Copperfield* speaking).[9] But why not? Perhaps our first impulse is to say, "Because we do not know who 'I' is and therefore have no way of verifying whether what he said did actually happen in real life."

It turns out, however, that this has little to do with how we actually make up our minds. We hardly ever have any independent means (history books, private records, and so on) of verifying what a man writes. More important, even our sense of whether his statements have "probability" on their side or not—even this, while it obviously makes a considerable difference to us, can never serve as an absolute distinction between truth-telling discourse, such as autobiography, and fictional discourse, such as a novel.

In the *Apologia Pro Vita Sua,* for example, there is a famous passage in which Newman, visiting Italy, begins to think he has a mission in life. An acquaintance urges him to visit Rome, but he replies

> with great gravity, "We have a work to do in England." I went down at once to Sicily, and the presentiment grew stronger. I struck into the middle of the island, and fell ill of a fever at Leonforte. My servant thought that I was dying, and begged for my last directions. I gave them, as he wished; but I said, "I shall not die." I repeated, "I shall not die, for I have not sinned against light, I have not sinned against light."[10]

After many tribulations, including being becalmed on an orange boat in the Straits of Bonifacio, he reaches his mother's house in England within hours of his brother's return from Persia and within five days of the great Assize Sermon that he later considers to have begun the Oxford Movement.

Here is an autobiographical passage that is apt to strike some readers (especially non-Catholic ones) as less "probable," more likely to have been shaped significantly by the author's purposeful imagination,[11] than almost anything in the highly probable and realistic *Robinson Crusoe.* "Probability" would therefore not work very well in the case of this one event if we wanted to decide whether the work it comes from were autobiography or fiction.

One might suppose that the examples above are too short, and have been unfairly lifted out of context, and that if we only had long enough passages it would be easy enough to decide whether a work is relating events likely or not to have happened in real life. "Probability" therefore still would work as a means of discriminating autobiography from fiction. This seems to make sense, for when we have a whole book in our hands we almost always know immediately whether we are to take it as autobiography or fiction (or something else); and we might be inclined to think the reason for this is that there is some larger "context" of words that does the trick even when short and perverse excerpts like the ones above do not.

Occasionally something like this does occur. We get a sense of an aggregate possibility built up word by word to make some large context that is saying "this book must be true" (or "fiction," or whatever)—one of those rare cases where we actually decide what genre a book belongs to as we go along. Much more often, however, the "probability" of the text has little or nothing to do with how we "take" a book. We decide on other grounds. A friend's or a critic's comment, or the library shelf, or the title page specifies a genre—specifies whether we are to take the book "as true" or not.

In the course of reading *David Copperfield* or the *Apologia,* for instance, no one decides whether the work is fiction or autobiography; he was "given" that before he began. In these cases we are not so much gathering a sense of probability *from* the text as engaging in a mental act that comes close to conferring a sense of probability *on* the text: that is, we have been told to make a certain assumption, and we therefore make it and are willing to persist in it even when a novelist says in the first person something temptingly like autobiography, or an autobiographer something suspiciously like fiction.[12] Thus a scheme for extracting sunbeams from cucumbers is perhaps a little less probable for being in a book titled *Gulliver's Travels* (Bk 3, chap. 5) than it might be in a histology journal. Even large contexts of "probability" in a work seldom, therefore, play much of a part in our decision whether to take the work "as true"; we usually have our minds made up for us at the start, and to some degree we give the author's "probabilities" the benefit of the doubt.[13]

What gives us this initial idea of the work? There are all sorts of answers, as we have seen. But here we come to the crucial point: for no matter how we got whatever impression we have of whether we ought to take the book "as true" or not—this of course says nothing about whether the book *is* "true." If the author himself, or the card catalogue, or something else gives us a sense that the book is to be considered as relating events that did really happen in life, we still know nothing whatever about the actual relation of the book to the "world."

What we do know is a supposed relation of the book to the author himself. That is, we almost never know the actual relation of literary events to what "happened, in real life," and on that basis decide whether we are reading fiction or autobiography; but we *do* know, or think we know, what the author *intends* the relation to be. We know that somebody (in McCarthy's case, apparently an editor at *Harper's*) is positing a relationship, such as "the author intends her work to be taken as fiction," or as "what really happened." The established assumption of what the author intends—that is everything.

The large-scale probability, the verisimilitude, of what an author says in his book therefore has to be considered merely a rhetorical strategy, like the title page and all sorts of other tricks of the trade, or to establish an intended genre for his work. The actual relation of the work to reality is problematical, and nothing more. Indeed, there is no reason why an autobiographer could not use some plausible event that never actually happened to help "declare" his work autobiographical; no reason why he should not reject an unbelievable event that *did* happen for fear of damaging his declaration; no reason why a novelist could not use that same event to help establish his own work as fiction.[14]

Therefore we have to concede at least in theory that absolutely anything a writer says can be "taken" as true or false if his declared or inferred intention is that we should do so. Whether or not David Copperfield's birth "happened, in real life" is irrelevant; we take it as fiction because we got an impression, in this case largely by means that have nothing to do with the "probability" of the text itself, of the author's intention (*"The Personal History, Adventures, Experiences, & Observations of David Copperfield The Younger of Blunderstone Rookery [Which He never meant to be Published on any Account]"*). The same would hold for a statement even if we had a sneaking suspicion it was a lie (such as a man's account of how he first appeared to his parents in a "green-silk Basket, such as neither Imagination nor authentic Spirits are wont to carry"), or a statement obviously a lie (such as the miraculous birth of Gargantua through the left ear of his mother).[15]

If the author or somebody else were to communicate to us, in some way we thought superseded what we could gather from the probability of these statements themselves, that we were to take them as true, we would have no choice but to try—no matter how much that taxed our credulity or patience. Our own appraisal of whether the statement is indeed true would present a problem: we would be likely to think that a man who sincerely claimed such things were autobiographical was either demented or insulting our intelligence. But in both cases the man would still be writing his autobiography no matter how pathetic or preposterous we were to consider him, and it.

The conclusion to be drawn reluctantly from this is that when we are determining which side of the great fact–fiction watershed a book belongs on, its conforming or not conforming to the facts of reality is irrelevant. We make our determination on the basis of the author's declared or inferred intention, and the probability or plausibility of what he says is merely one of many means at his disposal for establishing his intention. The genre "autobiography" seems therefore a matter of the author's intention, nothing more. As Shumaker puts it, the "determining consideration" is the "autobiographical intention." If the author "wishes to be understood as writing of himself and as setting down (so far as is humanly possible) nothing that is not literally and factually true, his work is autobiography. . . . If he wishes to be understood as writing imaginatively, it is something else."[16]

Thus the crucial distinction between "autobiography" and "fiction" is not to be found by comparing what a man writes to the "world outside"; the distinction seems to exist inside the writer's own head. So we now move from "outer" to "inner" matters. To begin with, what would make a writer decide to call his work autobiography, or fiction? Again, the answer seems simple. The writer, more than anybody else, is able to declare his writing one or the other on the basis of whether or not what he writes "happened, in real life."

But what does it mean to say that a related event (say, a man's eating hash for lunch) "happened, in real life"? Certainly not that the putative event got transferred into words. The event was, or occurred in, a mass existing in space and time; and all that words can do is make reference to such matters. So, obvious as it is, we have to let go at once of the comfortable idea that related events "happened."

Then the related words are a simulacrum, a verbal rendition, of the event. But what does that mean? Certainly not that the words somehow reproduce the entirety of the event. If McCarthy were to reproduce the top of the colonel's luncheon plate, there is still the hidden bottom, and also the hidden genealogy of the hash and the colonel. The autobiographer's words, like the novelist's and everybody else's, must artificially delimit, and therefore distort, the event. "Every limit is a beginning as well as an ending," as we are told in the last chapter of George Eliot's *Middlemarch.*

Thus related events have not "happened, in real life." The autobiographer, like every other relator of events fictive or "real," is engaged in a process of purposeful selection. "Autobiography means . . . selection in face of the endless complexity of life, selection of facts, distribution of emphases, choice of expression" (Pascal, p. 10).[17] In this respect an autobiographer is no different (except perhaps in degree?) from any other artist: he selects from what is somehow before him. A man "with absolutely truthful intention, amid the multitude of facts presented to him must needs select, and in selecting assert something of his own humour, something that comes not of the world without but a vision within."[18] So a man who chose to call his work autobiography or fiction solely on the basis of whether or not he made a purposeful selection from what he could have "put in" would be deluding somebody—himself or us.

Furthermore, the autobiographer, like the novelist, is apt to change the bits of reality he selects, and thus to fictionalize them. "Il faut changer la vie," as he knows.[19] Sometimes this happens without his knowing it. Like the rest of us, he has his quirks, anxieties, and mental defenses. He is on especially dangerous ground when, like McCarthy, he professes and apparently believes he is telling the truth. Such a grand purpose "has only to be proclaimed for it to become questionable. We know only too much about the inner censor, that acts so perfidiously because it acts automatically" (Pascal, p. 61).

But usually autobiographers know perfectly well what they are doing: like novelists, they are selecting and altering personal experience so as to create a structure of words that answers to an inner vision or purpose of some kind. As James Olney, the author of a fine, modern study of autobiography, notes: "One creates from moment to moment and continuously the reality to which one gives a metaphoric name and shape, and that shape is one's own shape."[20] McCarthy can claim that the details in "Artists in Uniform" are there because they "happened," but that cannot quite be the case. To say that a detail appears *because* it happened is not the same as saying it *did* happen. She says the former but must mean the latter. For there must have been other details that did happen during the experience but do not appear in the piece. Why are they absent?

Probably because they contributed nothing to, or even detracted from, the author's own sense of what her work ought to be like, a sense that could only be communicated, to the extent it could be communicated at all, by the details she chose (and created?). As for those absent details (the colonel's name, for instance), is it somehow untruthful to the event that they are absent? Certainly not. Because what the—shall we call it a story?—is being "true" to is not real life, but a conception an artist has derived from life, a conception that apparently got its initial impulse from life and then shaped itself as the writer began to cull from the experience (and alter, and create?) the details that would be the most adequate literary expression of that constantly changing conception (McCarthy comments on this herself).

Furthermore, the autobiographer, in trying to make his material answer to his vision, his purpose, his conception, is likely to be influenced by the demands of formal "convention" in much the same way a novelist is. As in the other arts, there are "conventions," there are "patterns" in autobiography.[21] "What one seeks in reading autobiography is not a date, a name, or a place, but a characteristic way of perceiving, of organizing . . ." (Olney, p. 37). The main (but not the only) strength of autobiography, like the main strength of all literature, is internal, in its "form," rather than in any correspondence to "fact," to "truth," to "what happened." The autobiographer therefore makes, in very general terms, the same kinds of aesthetic decisions, resulting in the same kind of aesthetic product, as any other literary artist. Indeed, we had better "speak of autobiography in the language of 'formal' literary criticism . . ." (Mandel, p. 218).[22]

So it seems that there is no reason for calling something "autobiography" that could not also be a reason for calling it "fiction." There is no way an author or anybody else could make a distinction between autobiography and fiction that would not crumble in his hands. There is nothing decisive in the relation of words to what happened, nothing decisive in the writer's own mental processes, nothing decisive in the organization of the words on the page—nothing that makes an essential distinction between the two kinds of writing. Regarded aesthetically, critically, "philosophically," the distinction so handy and familiar to us all breaks down. This is outrageous to common sense. But it seems an inexorable critical principle.

There is a saying, "Every artist writes his own autobiography"[23]— a saying that certainly seems borne out when a man in the preface to his own declared autobiography apologizes for the fact that some of the material has already appeared in his novels.[24] Why not leave it at that? All one's imaginative life is derived somehow from personal experience; all writing by some process of indirection "happened"—even magic casements in faery lands forlorn. "There is no Arte deliuered to mankinde that hath not the workes of Nature for his principall obiect . . ." (Sidney, p. 155). And likewise any account of what happened is bound to be colored by the imaginative life of the person telling the story. If we can accept all this, "autobiography" and "fiction" are—both autobiography and fiction; and we can look with perfect equanimity on the following title appearing in the "fiction" section of a magazine: "*Fredi & Shirl & the Kids: The Autobiography in Fables of Richard M. Elman, A Novel*, by Richard Elman."[25]

But no matter how much we question the rationale of distinguishing between "autobiography" and "fiction," no matter how thoroughly we convince ourselves that there is no essential difference between the two, the distinction does get made. Somehow an "intention" declares itself to us. So finally we must ask ourselves: what difference should that declared intention make in how we read the book?

McCarthy seems to be claiming it should make all the difference in the world. To the question, "What is the colonel's hash symbolic of?" she answers, "Symbolic of nothing, because what I wrote (*pace* my editors) is not fiction but autobiography." This seems to mean that a detail like the colonel's hash is potentially "symbolic" (in McCarthy's literary sense of referring to some idea beyond itself) or not symbolic according to whether the detail appears in something declared fiction or something declared autobiography. Does that make sense?

At first sight it does. If we get the idea that words on a page are an account of "what happened," we tend to assume with McCarthy that the related details are arrested in something like a literal stage; we assume that hash "means" simply hash. We assume that the details which exist in the work do so primarily, and probably exclusively, because they existed in a reality that is being recorded. Further inquiries into their "literary" meaning, inquiries into what ideas the details might have been contrived to "stand for" beyond themselves, seem almost, if not entirely, stultified. When Newman in the *Apologia* refers to the "orange boat" that carried him away from Sicily en route to his great work in England, not many readers (but see note 11 above) would want to look for any further significance whatever in this beyond its intrinsic interest as a vivid detail carried over from life.

But surely this attitude is irrational. Can we really put faith in whatever we can find out about an author's supposed state of mind at the time she wrote—whatever it is she now says she wrote? If she seems to recall that the events in her work somehow derived from what she remembers as "life" to an extent that, even considering omissions, revisions, and other transmogrifications, she now for whatever reason considers significant, does any declaration of this cut off the possibility of the kinds of symbolic meaning there would be if the work had somehow got itself classified as "fiction"?

Are we supposed to believe that the word "hash" on a page of *Harper's* has potentially a symbolic meaning for about a year, then changes and no longer has that potential meaning because the author (or suppose someone else) wrote in a book or a letter or whispered to a friend that she remembers there existed real hash on a plate that she now identifies with the words "corned-beef hash" on page 47 of the March 1953 *Harper's?* Was there anything else on the real colonel's plate, a pickled pear omitted in the story but existing in reality, so that McCarthy is now in the state of reporting real hash that happened on an otherwise fictive plate? Can the rest of the plate be "symbolic" but not the hash? What if the station restaurant reports in and says it was lamb hash—can we go back to symbolizing the word in the story? In *David Copperfield* should we fence off whatever details seem to have a more direct relation than the others to Dickens's own life, and not consider them as "literature?"

In short, should a work have a different status, a different ontology as an artifact, according to somebody's declared guess concerning the relation of "what happened" to words on a page? Can there be any reasonable answer but No?

In the archangel Michael's words to Adam, "Let us descend now therefore from this top / Of Speculation." Suppose we are convinced to our satisfaction (or, even better, to our dissatisfaction) that there are no grounds for an essential distinction between autobiography and fiction. Nevertheless, the two genres do exist, and we are always encountering and relying on the distinction.

How is it that we can convince ourselves there is nothing intrinsic in works that allows us to make an absolute distinction between "fact" and "fiction," yet use the absolute distinction all the time? The answer lies in the constitution of the mind. Any discrete thing external to the mind can be perceived as a *Ding an sich,* or it can be perceived "referentially," as referring to something beyond itself. It can be taken to refer only to itself, to "be," no matter how tempting its relation to a context beyond it; or it can be taken to refer to some context outside itself, no matter how oblique or stylized the reference. This is true of both a soup can and a book. We can regard either as a self-referential object, hence "aesthetically," as "art." And we can regard either as a referent, a "sign" referring to, and hence making some kind of "statement" about, a context outside itself: "Open me and there is something to eat"; "Open me and see what happened, in real life."

There is no way of breaking down this distinction, although a formulation that takes into account the obvious complexities is not easy to arrive at. A piece of sculpture can have "reference," and can even be conceived as making a statement useful in our lives; and a useful object like a can or a hammer may have "beauty." But that is merely a way of saying we can be "aware" of the mode of apprehension we are not committed to, and may even have been committed to it rather than the other a moment ago. Indeed, we are always aware at least faintly of the possibility of apprehending a thing the other way, the "truth" of statues and the "beauty" of hammers being a tension between whichever mode we commit ourselves to and the attractive possibilities of the other. In a context similar to mine, Edmund Husserl refers to this tension as "perceptual consciousness resolved in conflict."[26]

But the distinction itself is absolute, as Husserl emphasizes. An arrow has got to point in a direction. A thing has got to be conceived rather as referring or not referring: referring outward or inward. To imagine both at once is actually to imagine one followed by the other. There is no mentally tenable third mode that combines the two: "Take this thing as making a statement about 'life' but as merely existing without making any such statement." Hence "truth" and "beauty" always seem to fall into some kind of dialectical relationship.

Therefore that Great Divide between "fact" and "fiction" does not exist on the literary map at all; surely one can demonstrate that it does not. But such a Divide does exist nevertheless, in the very constitution of the mind, no matter what the mind is contemplating. Whether or not there is any absolute distinction in "kind" between a soup can and a Brancusi, between directions on an aspirin bottle and the *Apologia,* there is an absolute distinction between the two ways the mind can contemplate whatever is before it. Thus in the very constitution of the mind we have the rudiments of genre. All texts tend to be a conflation of fact and fiction. But the intending or inferring mind has *got* to declare for one or the other in spite of the unimpeachable truth that such a distinction does not "exist" in the texts themselves.

What we have now is a continuum of texts, concerning which the author's or somebody else's mind has made a reductive, artificial distinction: "Regardless of what this really is, consider it fact (or fiction)." Indeed, the various kinds of printed words arrange themselves in a shape like a horseshoe magnet. At one pole are texts such as the word "PUSH" on a door. Here the sign-maker's "intention" (probably gathered from the "context," a door) declares unambiguously that the sign be considered as "fact," as a reference to a state of affairs concerning how to open the door. A man would be hard put to conceive the sign "aesthetically" (for its onomatopoeia?).

We now move up the magnet away from the pole, through more and more complex organizations of "signs" that nevertheless manage to declare themselves "referential" (directions on an aspirin bottle, an article in a physics journal, and a history journal), until we get to texts like Darwin's account of the voyage of the *Beagle* and Gibbon's *Decline and Fall*. About here, about the place where we begin to talk about science and history books as having "style," our two modes begin to generate a little tension between themselves. The books are still managing to declare themselves "referential," but we begin to think that with a little perverseness of will we might change our apprehension to the aesthetic mode—and think of Darwin's ship as similar to Coleridge's in "The Rime of the Ancient Mariner."

Move a little further along the curve of the magnet and we are in a middle area (what would be the handle of the magnet) where the "intention" becomes more and more problematical, where apprehending the work in the mode of "truth" seems more or less balanced by the possibility of apprehending it in the mode of "beauty." Somewhere in this area, close to the center of the handle, we come to autobiography. Hence the problem, what is its intention, how are we to "take" it: as reference, as a type of history (the Library of Congress solution); or as "art," in which case we try to block out reference and to experience the work as poetry or fiction. Just to one side of center would probably be an autobiography like Newman's *Apologia,* where the context, the "intention," seems to be telling us to read for "what happened" (but presented with what style, what artistic purposefulness!); but where the context, the intention, seems also to be telling us that we *could* take the whole book aesthetically, as "fiction" (but with what truth to the author's own life!).

To the other side of center would be *Sartor Resartus.* As we move down this arm of the magnet, the tension between the two modes gradually relaxes the other way. We approach the "fiction" shelf in the library, which unambiguously declares for the aesthetic mode. Then poetry, which may have its theoretical terminus in whatever esoteric imagism one could imagine where the words, those signs pointing toward the world, have been so transformed by context, so bent inward, that external reference could be conceived as minimal. We are now in those faery lands forlorn, where it is far from easy to find aspirin-bottle "statements," cryptograms in the trees, sermons in stones. A little further and we approach the other *ultima Thule,* where direct external "reference" has (almost?) been locked out, and meaning is created solely by nonverbal signs moving about in a closed, self-referential context—mathematics. Thus "PUSH" and "$x = \frac{1}{2}(e^x + e^{-x})$" are poles apart, and in the precision and absoluteness of their mode of reference almost touching. Extremes meet.

Now, imagine an ideally "referential" mind confronted with our magnet. Such a mind would be comfortable at the "PUSH" pole and would be under increasing tension as it traveled toward the handle and the other pole. This mind would read Darwin's voyage for "truth," what happened, but would also be committed to reading *David Copperfield* as much as possible the same way (for moral apothegms that refer to life? for what "happened" to Dickens?). Likewise Coleridge's poem (for a moral, like "He prayeth best, who loveth best / All things both great and small"? for what happened to a real mariner? or to Coleridge?), and likewise the hyperbolic cosine of x (here the mind boggles: a symbolic "statement" to apply to one's life? a symbolic analogue of real events?).

There is of course no such ideally "referential" mind. But as for an ideally "aesthetic" one—enter the critic. To him as critic everything from doorsigns and soup cans to cosines is to be apprehended in the "aesthetic" mode, and thus framed in a way that holds off any direct reference to life. Put more delicately, he has to see how the "sign" does indeed "refer" to reality in the act of deviating from it in whatever way allows an aesthetic apprehension (otherwise the "art" would collapse back into the real it "imitates"); but his awareness of reference is supposed to remain suspended in that precarious state. This, regardless of what the sign really "is," regardless of which intention it seems to be declaring for itself, or which of the two modes he would personally like to bring to bear on it. He may personally want to find the cloakroom, or be hungry, or be curious about Dickens's personal life; but to him as a critic all signs must be aesthetic signs.

To him as a critic a soup can is art. The can is now "like" a real can, but is an aesthetic object; he is not supposed to want to eat the contents, or judge the art according to whether he likes black bean more than asparagus. Likewise, he reads Coleridge's poem and knows that the ship is like a real ship (otherwise it could not be given the name) but is a poetic ship; it is not supposed to matter much to him as a critic whether the ship really existed, whether the adventures on it "happened, in real life."

And so to the critic as critic all organizations of words are fiction! Fiction in its various forms. He looks in the glossary of *Anatomy of Criticism* and finds that "fiction" is another word for printed litera-ture.[27] He finds nothing disturbing in this. He reads a first-aid manual "aesthetically," as "fiction," keeping his awareness of "refer-ence" in that suspended state, and concentrating on structure, imag-ery, style.[28] He does the same with Dickens—and Henry Adams.

Yes, he reads autobiography as fiction. As for the *Apologia,* it "begins in the practical world and ends, as it were, framed like a novel. . . . it survives as literature because it can sustain the con-templative and unpragmatic reading." Likewise, "anyone who at-tempts to teach Mill's *Autobiography* . . . quickly realizes that he is engaged in discussing a response to experience that differs very little if at all from the kind of thing he has been accustomed to think of as the peculiar activity of poets and novelists."[29]

The critic tries to assume that over every autobiography is hung a sign that says, "The opinions expressed here are not necessarily those of the management." If McCarthy says that the hash on the page corresponds to real hash in a restaurant, the critic does not care—or tries not to. To him *hash* is a word in an ultimately self-referring structure of words, and is therefore as potentially "sym-bolic" as any other word in that structure. Where the words came from does not matter; what to make of them is strictly his business, not McCarthy's. The job he took on when he dedicated himself to the aesthetic apprehension rather than the other was to try to see autobiography and everything else as cut off from any direct refer-ence to what happened, in real life. "The secret to experiencing non-fiction prose as art lies not in the text but in ourselves: to the degree we can turn off our normal reality-testing, we can be drawn into non-fiction as a literary experience like our experience of poetry or fiction."[30]

That makes autobiography a special problem for him. Nobody ever expects him to read a doorsign or Darwin in a state of mind that suspends his awareness of reference; but autobiography, espe-cially in recent years, is at least visible on his horizon. This taxes his aesthetic apprehension to the limit, for, as we have seen, autobiography exists in that area of our conceptual model where the possibilities of the two modes of apprehension are about balanced. Hence an area where words can be read either way, where "symbols" can be conceived as pointing either outward or inward (or outward-then-inward).

But not both ways at once. A crude and arbitrary decision has *got* to be made; and the critic has opted for beauty. Thus when he reads autobiography he cuts himself off from "reference" he knows is there, but that cannot be the final issue of his thoughts. He has this problem with all texts to some extent; but he has the problem more in the case of autobiography than any other work with which he habitually deals. All texts are fact and fiction, but autobiography most of all. Hence his mind is under great tension. Can he ignore the potential truth of autobiography? Can he suspend his "reality-testing"? Can he resist reading the sign but not opening the door? Can he resist reading Newman for "what happened"? That is why for criticism as a discipline autobiography will always be a problem, and why the question of the colonel's hash can be settled theoretically—but never quite be settled.

NOTES

1. *Harper's Magazine,* February 1954, pp. 68–75. Both articles are reprinted in this volume.

2. Ed. J. Shawcross, 2 vols. (Oxford: Oxford University Press, 1907), II, p. 10.

3. Wellek and Warren, *Theory of Literature* (New York: Harcourt Brace, 1949), p. 11; Brooks, *The Well Wrought Urn* (New York: Harcourt Brace, 1947), p. 210; Frye, *Anatomy of Criticism* (Princeton: Princeton University Press, 1957), p. 74; Richards, *Principles of Literary Criticism* (New York: Harcourt, Brace, 1924), p. 266; Rader, "Literary Form in Factual Narrative: The Example of Boswell's *Johnson,*" in *Essays in Eighteenth-Century Biography,* Philip B. Daghlian, ed. (Bloomington: Indiana University Press, 1968), p. 3.

4. Frye uses this example (p. 303). I note that the bibliography in Roy Pascal's *Design and Truth in Autobiography* (Cambridge, Mass.: Harvard University Press, 1960) lists Somerset Maugham's *Of Human Bondage* as an autobiography: Maugham himself called it an "autobiographical novel" (*The Summing Up* [London: W. Heinemann, 1938], p. 196); the Library of Congress system classifies it as fiction (PR).

5. See John Phillip Immroth, *A Guide to Library of Congress Classification* (Rochester: N.Y. Libraries Unlimited, 1968), p. 180.

6. Shumaker, *English Autobiography* (Berkeley: University of California Press, 1954), p. 101; Mandel, "The Autobiographer's Art," *JAAC* 27: 219 (1968); Ham-

burger, *The Logic of Literature,* Marilynn J. Rose, trans. 2nd ed. (Bloomington: Indiana University Press, 1973), pp. 312, 328–29.

7. Following is a sampler of statements to this effect, none of them easily comprehensible. Georg Misch, *A History of Autobiography in Antiquity,* 2 vols. (London: Routledge and Paul, 1950), I, p. 11: "in general, the spirit brooding over the recollected material is the truest and most real element in an autobiography." Georges Gusdorf, "Conditions et limites de l'autobiographie," in *Formen der Selbstdarstellung,* Günter Reichenkron and Erich Haase, eds. (Berlin: Duncker and Humblot, 1956): "l'autobiographie est une seconde lecture de l'expérience, et plus vraie que la première, puisqu'elle en est la prise de conscience" (p. 114); "dans le cas de l'autobiographie, la vérité des faits apparaît subordonnée à la vérité de l'homme . . ." (p. 118).

8. *An Apology for Poetry,* in *Elizabethan Critical Essays,* G. Gregory Smith, ed. 2 vols. (1904; rpt. London: Clarendon Press, 1964), I. p. 185.

9. Chapter 1 of *David Copperfield,* edited for my own purposes.

10. John Henry Newman, *Apologia Pro Vita Sua,* Martin J. Svaglic, ed. (Oxford: Clarendon Press, 1967), p. 43.

11. George Levine, *The Boundaries of Fiction* (Princeton: Princeton University Press, 1968), pp. 246–50, makes much the same point about this passage.

12. For a discussion of how we stubbornly make the details of a text conform to our preconceived idea of the genre of the text, see E. D. Hirsch, Jr., *Validity in Interpretation* (New Haven: Yale University Press, 1967), pp. 164–69. "It is very difficult to dislodge or relinquish one's own genre idea, since that idea seems so totally adequate to the text" (p. 166).

13. There is a similar comment of Ludwig Wittgenstein's on the difference "context" makes: "I describe a psychological experiment: the apparatus, the questions of the experimenter, the actions and replies of the subject—and then I say that it is a scene in a play.—Now everything is different" (*Philosophical Investigations,* G. E. M. Anscombe, trans. 2nd ed. [New York: B. Blackwell, 1958], p. 180e).

14. But Newman makes an interesting comment on this: "Miss Edgeworth sometimes apologizes for certain incident [*sic*] in her tales, by stating they took place 'by one of those strange chances which occur in life, but seem incredible when found in writing.' Such an excuse evinces a misconception of the principle of fiction, which, being the perfection of the actual, prohibits the introduction of any such anomalies of experience" ("Poetry, with Reference to Aristotle's Poetics," *Essays Critical and Historical* [London: B. M. Pickering, 1901], I, pp. 13–14).

15. The first example is the arrival of the baby Teufelsdröckh, *Sartor Resartus,* Bk. 2, chap. 1; the second, *Gargantua,* Bk. 1, Chap. 6.

16. I have edited a (minor?) proviso out of the quotation, p. 105.

17. Mandel makes the same point (p. 218).

18. Walter Pater, "Style," *Appreciations,* in *Works,* New Library Ed., 10 vols. (London: Macmillan, 1910), V, p. 9.

19. Michel Butor, "Une Autobiographie dialectique," *Répertoire,* I (Paris: Editions de minuit, 1960), p. 262.

20. *Metaphors of Self: The Meaning of Autobiography* (Princeton: Princeton University Press, 1972), p. 34. Pascal makes a similar comment (p. 72).

21. See Pascal, p. 2; Levine, p. 170: "One of the distinctive characteristics of autobiography . . . is the imposition of a pattern . . ."

22. Gusdorf also has a relevant comment: "Il faut donc admettre une sorte de renversement de perspective, et renoncer à considérer l'autobiographie à la manière d'une biographie objective, régie par les seules exigences du genre historique. Toute autobiographie est une oeuvre d'art" (p. 120).

23. Cited by H. N. Wethered, *The Curious Art of Autobiography* (London: C. Johnson, 1956), p. 2.

24. Nikolai Gubsky, *Angry Dust: An Autobiography* (New York: Oxford University Press, 1937), p. ix.

25. The equanimity of the reviewer was a little less than perfect: "The first question this book raises is whether the 'Richard Elman' given as the author is any different from the 'Richard M. Elman' in the subtitle . . ." (*New Yorker,* August 5, 1972, p. 82).

26. *Logical Investigations,* J. N. Findlay, trans. 2 vols. (London: Routledge and Kegan Paul, 1970), II, p. 610. "Two perceptual interpretations . . . of a thing interpenetrate. . . . And they interpenetrate in conflicting fashion, so that our observation wanders from one to another . . . each barring the other from existence" (II, p. 610).

27. See also Frye, p. 303; and R. G. Collingwood, *The Principles of Art* (Oxford: Oxford University Press, 1950), p. 285.

28. See the critic Geoffrey Tillotson at work on a paragraph titled "Fainting," in "Matthew Arnold's Prose: Theory and Practice," in *The Art of Victorian Prose,* George Levine and William Madden, eds. (New York: Oxford University Press, 1968), pp. 75–76.

29. Introduction to *The Art of Victorian Prose,* pp. xix, vii.

30. Norman Holland, "Prose and Minds: A Psychoanalytic Approach to Non-Fiction," in *The Art of Victorian Prose,* p. 322.

Selected Bibliography

This bibliography is more suggestive than it is complete. With a few exceptions, autobiographies and memoirs have been excluded. Most of the works were written during the twentieth century, although a few nineteenth century classics are included. Often only a few works by a particular author are mentioned. For example, John McPhee has more than twenty books in print, but only a couple are mentioned here. This list is intended as a starting point for further reading and research.

Original works come first, arranged in three categories according to the date when they were written, followed by works focused on literary theory and research, and then historical sources and studies.

ORIGINAL WORKS

Before 1915

Ade, George. *Chicago Stories.* Edited by Franklin J. Meine; illustrated by John T. McCutcheon and others. Chicago: Henry Regnery, 1963. Selected journalism from early in the century.

Baker, Ray Stannard. *Following the Color Line.* New York: Harper Torchbooks, 1964. Originally published by Doubleday, Page & Co. in 1908.

Berkman, Alexander. *Prison Memoirs of an Anarchist.* Introduction by Hutchins Hapgood; new introduction by Paul Goodman. New York: Schocken Books, 1970. Originally published 1912.

Borrow, George. *The Bible in Spain.* New York: G. P. Putnam's Sons, 1899.

————. *Lavengro: The Scholar, the Gypsy, the Priest.* New York: Macmillan, 1927.

————. *The Zincali: An Account of the Gypsies of Spain.* London: John Murray, 1923.

Cahan, Abraham. *Grandma Never Lived in America: The New Journalism of Abraham Cahan.* Edited with an introduction by Moses Rischin. Bloomington: Indiana University Press, 1985.

Crane, Stephen. *Tales, Sketches, and Reports,* Vol. 7 of *The Works of Stephen Crane,* edited by Fredson Bowers. Charlottesville: University Press of Virginia, 1973.

Davis, Richard Harding. *The Red Cross Girl.* New York: Charles Scribner's Sons, 1916.

————. *A Year from a Reporter's Note-Book.* New York: Harper & Brothers, 1897.

Dreiser, Theodore. *Selected Magazine Articles of Theodore Dreiser; Life and Art in the American 1890s.* Edited, with an introduction and notes, by Yoshinobu Hakutani. Rutherford, Madison, Teaneck: Fairleigh Dickinson University Press, 1985.

Du Bois, W. E. B. *The Souls of Black Folk,* edited by Saunders Redding. New York: Fawcett, 1961. Originally published in 1903.

————. *W. E. B. DuBois: Writings,* edited by Nathan Huggins. New York: The Library of America, 1986.

Dunne, Finley Peter. *Mr. Dooley at His Best.* Edited by Elmer Ellis. New York: Scribner's, 1943.

Durland, Kellogg. *The Red Reign: The True Story of an Adventurous Year in Russia.* New York: The Century Co., 1907.

Friedman, I. K. *The Lucky Number.* Chicago: Way and Williams, 1896.

Flynt, Josiah. *Tramping with Tramps; Studies and Sketches of Vagabond Life.* New York: The Century Co., 1899.

Hapgood, Hutchins. *An Anarchist Woman.* New York: Duffield & Co., 1909.

————. *The Spirit of the Ghetto: Studies of the Jewish Quarter in New York.* Introduction by Moses Rischin. Cambridge: Harvard University Press, 1967. Originally published in 1902.

————. *The Spirit of Labor.* New York: Duffield & Co., 1907.

————. *Types from City Streets.* New York and London: Funk & Wagnalls Co., 1910. Magazine articles and sketches.

Hearn, Lafcadio. *Fantastics and Other Fancies.* Boston and New York: Houghton Mifflin, 1914. New Orleans newspaper articles, 1879–1884.

London, Jack. *Jack London on the Road; The Tramp Diary and Other Hobo Writings.* Edited by Richard W. Etulain. Logan, Utah: Utah State University Press, 1979.

————. *The People of the Abyss.* New York: Grosset & Dunlap, 1903.

————. *The Road.* New York: The Macmillan Co., 1907.

Reed, John. *Adventures of a Young Man; Short Stories from Life.* San Francisco: City Lights Books, 1975.

————. *The Education of John Reed.* With an introductory essay by John Stuart. New York: International Publishers, 1955.

————. *Insurgent Mexico.* New York: International Publishers, 1969. First published 1914.

Steffens, Lincoln. *The Shame of the Cities.* New York: McClure, Phillips and Co., 1904.

Turgenev, Ivan. *A Sportsman's Notebook.* New York: The Viking Press, 1950. Originally published 1852.

Twain, Mark. *The Complete Humorous Sketches and Tales of Mark Twain.* Edited, and with an introduction, by Charles Neider. Garden City, N.Y.: Doubleday & Co., 1961.

————. *Early Tales & Sketches.* Vols. 1 (1851–1864) and 2 (1864–1865). Edited by Edgar Marquess Branch and Robert H. Hirst with the assistance of Harriet Elinor Smith. Berkeley, Los Angeles: University of California Press, 1979.

————. *Life on the Mississippi.* New York: New American Library, 1980. Originally published 1883.

Wyckoff, Walter A. *A Day with a Tramp, and Other Days.* New York: Benjamin Blom Inc., 1971. Originally published 1901. Reports his experiences as a day laborer and tramp in 1891. Wyckoff later taught at Princeton.

Before 1960

Adamic, Louis. *My America 1928–1938.* New York: Harper & Brothers, 1938. Depression-era reporting and profiles.

Agee, James. *James Agee: Selected Journalism.* Edited, with an introduction, by Paul Ashdown. Knoxville: University of Tennessee Press, 1985.

————. *Let Us Now Praise Famous Men.* Photos by Walker Evans, with an introduction by John Hersey. Boston: Houghton Mifflin, 1988. Originally published in 1941. The most recent of several editions.

Berger, Meyer. *The Eight Million; Journal of a New York Correspondent.* New York: Columbia University Press, 1983. Originally published by Simon and Schuster, 1942.

Bolitho, William. *Camera Obscura.* New York: Simon and Schuster, 1930.

Capote, Truman. *The Muses Are Heard.* New York: Random House, 1956.

Dorr, Rheta Childe. *Inside the Russian Revolution.* New York: Macmillan Co., 1918. Reprinted by Arno Press in 1970.

Dos Passos, John. *In All Countries.* New York: Harcourt, Brace and Co., 1934. Collected articles about Russia, Mexico, and the United States.

————. *Orient Express.* New York: Harper & Brothers, 1922, 1927.

————. *Rosinante to the Road Again.* New York: George H. Doran Co., 1922.

Dreiser, Theodore. *The Color of a Great City.* New York: Boni and Liveright, 1923.

————. *Theodore Dreiser: A Selection of Uncollected Prose.* Detroit: Wayne State University Press, 1977.

Flanner, Janet. *Paris Journal 1944–1965.* Edited by William Shawn. New York: Atheneum, 1965.

————. *Paris Was Yesterday (1925–1939).* New York: Popular Library, 1972.

Hecht, Ben. *1001 Afternoons in Chicago.* Chicago: Pascal Covici, 1922.

Hemingway, Ernest. *By-Line: Ernest Hemingway.* Edited by William White. New York: Scribner's, 1967.

————. *The Dangerous Summer.* New York: Scribner's, 1985. Originally published 1960.

————. *Dateline: Toronto; The Complete Toronto Star Dispatches, 1920–1924.* Edited by William White. New York: Charles Scribner's Sons, 1985.

————. *Death in the Afternoon.* New York: Scribner's, 1932.

————. *Green Hills of Africa.* New York: Scribner's, 1935.

————. *A Moveable Feast.* New York: Scribner's, 1964.

Hersey, John. *Here to Stay.* New York: Alfred A. Knopf, 1963. Collected stories from World War II.

————. *Hiroshima.* New York: Alfred A. Knopf, 1946.

Liebling, A. J. *Back Where I Came From.* New York: Sheridan House, 1938.

————. *Liebling Abroad.* Introduction by Raymond Sokolov. Playboy Press, 1981. Contains: *The Road Back to Paris, Mollie & Other War Pieces, Normandy Revisited, Between Meals: An Appetite for Paris.*

————. *Liebling at Home.* Introduction by Herbert Mitgang. Wideview Books, 1982. Contains: *The Telephone Booth Indian; Chicago: The Second City; The Honest Rainmaker; The Earl of Louisiana;* and *The Jolity Building.*

————. *The Sweet Science.* New York: The Viking Press, 1956.

McCarthy, Mary. *On the Contrary.* New York: Farrar, Straus and Cudahy, 1961.

McKelway, St. Clair. *True Tales from the Annals of Crime and Rascality.* New York: Random House, 1951.

McNulty, John. *A Man Gets Around.* Boston: Little, Brown and Co., 1951.

————. *Third Avenue, New York.* Boston: Little, Brown and Co., 1946.

————. *The World of John McNulty.* With an appreciation by James Thurber. Garden City, N.Y.: Doubleday & Co., 1957

Mitchell, Joseph. *The Bottom of the Harbor.* Boston: Little, Brown and Company, 1960.

————. *McSorley's Wonderful Saloon.* New York: Duell, Sloan and Pearce, 1943.

————. "The Mohawks in High Steel," in *Apologies to the Iroquois,* with Edmund Wilson. New York: Farrar, Straus, 1960.

————. *My Ears Are Bent.* New York: Sheridan House, 1938.

————. *Old Mr. Flood.* New York: Duell, Sloan and Pearce, 1948.

The New Republic Anthology 1915–1935. Groff Conklin, ed. New York: Dodge Publishing, 1936.

The New Yorker Book of War Pieces. New York: Reynal & Hitchcock, 1947. Reissued 1989.

Orwell, George. *Down and Out in Paris and London.* London: Gollancz, 1933.

————. *Homage to Catalonia.* London: Secker & Warburg, 1938.

————. *The Road to Wigan Pier.* London: Gollancz, 1937.

Paul, Elliot. *The Life and Death of a Spanish Town.* New York: Random House, 1937. Reprinted by Greenwood Press, 1971.

————. *The Last Time I Saw Paris.* New York: Random House, 1942.

Reed, John. *Ten Days That Shook the World.* Foreword by V. I. Lenin. New York: Vintage Books, 1960. First published 1919.

————. *The War in Eastern Europe.* New York: Charles Scribner's Sons, 1918.

Ross, Lillian. *Picture: John Huston, M.G.M., and the Making of The Red Badge of Courage.* New York: Limelight Editions, 1984. First published in 1952.

————. *Reporting.* New York: Dodd, Mead & Co., 1964, 1981.

————. *Takes: Stories from The Talk of the Town.* New York: Congdon & Weed, Inc., 1983.

————. *Talk Stories.* New York: Simon and Schuster, 1966.

Shirer, William L. *Berlin Diary.* New York: Alfred A. Knopf, 1941.

Steinbeck, John. *The Grapes of Wrath: Text and Criticism.* Edited by Peter Lisca. New York: The Viking Press, 1972.

————. *The Log from the Sea of Cortez.* With E. F. Ricketts. New York: The Viking Press, 1951.

————. *Once There Was a War.* New York: The Viking Press, 1958. First published in 1943.

————. *A Russian Journal.* With pictures by Robert Capa. New York: The Viking Press, 1948.

————. *Travels with Charley; In Search of America.* New York: The Viking Press, 1962.

Swados, Harvey, editor. *The American Writer and the Great Depression.* Indianapolis: Bobbs-Merrill Educational Publishing, 1966.

West, Rebecca. *Black Lamb and Grey Falcon; A Journey through Yugoslavia.* New York: The Viking Press, 1941.

————. *Rebecca West: A Celebration.* Selected from her writings, with a critical introduction by Samuel Hynes. New York: Penguin Books, 1978.

Wilson, Edmund. *The American Earthquake.* Garden City, N.Y.: Doubleday & Co., 1958.

White, E. B. *Essays of E. B. White.* New York: Harper Colophon Books, 1977.

————. *Here Is New York.* New York: Harper & Brothers, 1949.

————. *The Second Tree from the Corner.* New York: Harper & Row, 1965.

Since 1960

Anzaldúa, Gloria. *Borderlands/La Frontera; The New Mestiza.* San Francisco: Spinsters/Aunt Lute Book Co., 1987.

Anzaldúa, Gloria, and Cherríe Moraga. *This Bridge Called My Back: Writing by Radical Women of Color.* New York: Kitchen Table: Women of Color Press, 1983.

Bedford, Sybille. *The Faces of Justice; A Traveller's Report.* New York: Simon and Schuster, 1961.

Capote, Truman. *In Cold Blood.* New York: Random House, 1965.

Conover, Ted. *Coyotes: A Journey through the Secret World of America's Illegal Aliens.* New York: Vintage Books, 1987.

Davidson, Sara. *Loose Change: Three Women of the Sixties.* Garden City, N.Y.: Doubleday, 1977.

Didion, Joan. *Miami.* New York: Simon and Schuster, 1987.

————. *Salvador.* New York: Simon and Schuster, 1983.

————. *Slouching Towards Bethlehem.* New York: Farrar, Straus & Giroux, 1968.

———. *The White Album.* New York: Simon and Schuster, 1979.

Dunne, John Gregory. *Vegas: A Memoir of a Dark Season.* New York: Random House, 1974.

Ehrlich, Gretel. *The Solace of Open Spaces.* New York: Viking, 1985.

Fitzgerald, Frances. *Cities on a Hill: A Journey through Contemporary American Cultures.* New York: Simon and Schuster, 1986.

Frady, Marshall. *Southerners: A Journalist's Odyssey.* New York: New American Library, 1980.

Frazer, Ian. *Great Plains.* New York: Farrar, Straus & Giroux, 1989.

Herr, Michael, *Dispatches.* New York: Alfred A. Knopf, 1977.

Hersey, John. *Blues.* New York: Alfred A. Knopf, 1987.

Hoagland, Edward. *The Edward Hoagland Reader.* Edited and with an introduction by Geoffrey Wolff. New York: Random House, 1979.

Hubbell, Sue. *A Book of Bees.* New York: Random House, 1988.

Kidder, Tracy. *Among Schoolchildren.* Boston: Houghton Mifflin, 1989.

———. *The Soul of a New Machine.* Boston: Little, Brown and Co., 1981.

Kramer, Jane. *Europeans.* New York: Farrar, Straus & Giroux, 1988.

———. *The Last Cowboy.* New York: Harper & Row, 1977.

Kramer, Mark. *Invasive Procedures: A Year in the World of Two Surgeons.* New York: Harper & Row, 1983.

Lernoux, Penny. *Cry of the People.* Garden City, N.Y.: Doubleday, 1982.

Levine, Richard M. *Bad Blood: A Family Murder in Marin County.* New York: Random House, 1982.

Lopez, Barry. *Arctic Dreams: Imagination and Desire in a Northern Landscape.* New York: Charles Scribner's Sons, 1986.

Lukas, J. Anthony. *Common Ground: A Turbulent Decade in the Lives of Three American Families.* New York: Alfred A. Knopf, 1985.

Mailer, Norman. *The Armies of the Night: History as a Novel; The Novel as History.* New York: New American Library, 1968.

———. *The Executioner's Song.* Boston: Little, Brown & Co., 1979.

———. *Miami and the Siege of Chicago.* New York: New American Library, 1968.

———. *Some Honorable Men: Political Conventions 1960–1972.* Boston: Little, Brown & Co., 1976. Contains "Superman Comes to the Supermarket."

Matthiessen, Peter. *The Snow Leopard.* New York: Viking Press, 1978.

McCarthy, Mary. *Vietnam.* New York: Harcourt, Brace & World, 1967.

———. *Hanoi.* London: Weidenfeld and Nicholson, 1968.

McKelway, St. Clair. *The Big Little Man from Brooklyn.* Boston: Houghton Mifflin Co., 1969

———. *The Edinburgh Caper; A One-Man International Plot.* New York: Holt, Rinehart and Winston, 1962.

McPhee, John. *The Control of Nature.* New York: Farrar, Straus & Giroux, 1989.

———. *Encounters with the Archdruid.* New York: Farrar, Straus, & Giroux, 1971.

———. *The John McPhee Reader.* Edited, with an introduction, by William Howarth. New York: Farrar, Straus, & Giroux, 1976.

Mitchell, Joseph. *Joe Gould's Secret.* New York: The Viking Press, 1965.

Naipaul, V. S. *A Turn in the South.* New York: Alfred A. Knopf, 1989.

Olsen, Tillie. *Silences.* New York: Delta/Seymour Lawrence, 1989. Originally published in 1978.

Preston, Richard. *First Light: The Search for the Edge of the Universe.* New York: Atlantic Monthly Press, 1987.

Rhodes, Richard. *Farm: A Year in the Life of an American Farmer.* New York: Simon and Schuster, 1989.

———. *The Making of the Atomic Bomb.* New York: Simon and Schuster, 1986.

Sheehan, Susan. *Kate Quinton's Days.* Boston: Houghton Mifflin, 1984.

Sims, Norman, ed. *The Literary Journalists.* New York: Ballantine Books, 1984.

Singer, Mark. *Funny Money.* New York: Alfred A. Knopf, 1985.

Snow, Edgar. *Red China Today.* New York: Random House, 1970. Orignally published as *The Other Side of the River* in 1962.

Talese, Gay. *Honor Thy Father.* New York: Fawcett-World, 1972.

———. *The Kingdom and The Power.* New York: World Publishing Co., 1966.

———. *Thy Neighbor's Wife.* Garden City, N.Y.: Doubleday, 1980.

Thompson, Hunter. *Fear and Loathing in Las Vegas: A Savage Journey to the Heart of the American Dream.* New York: Popular Library, 1971.

———. *The Great Shark Hunt.* New York: Fawcett, 1979.

Trillin, Calvin. *Killings.* New York: Ticknor & Fields, 1984.

———. *U.S. Journal.* New York: Dutton, 1971

Warnock, John. *Representing Reality: Readings in Literary Nonfiction.* New York: St. Martin's Press, 1989.

White, Theodore H. *The Making of the President 1960.* New York: Atheneum, 1961.

Wicker, Tom. *A Time to Die.* New York: Quadrangle, 1975.

Wideman, John Edgar. *Brothers & Keepers.* New York: Holt Rinehart and Winston, 1984.

Wilkinson, Alec. *Midnights: A Year with the Wellfleet Police.* New York: Random House, 1982.

Wolfe, Tom. *The Electric Kool-Aid Acid Test.* New York: Farrar, Straus & Giroux, 1970.

————. *The Kandy-Kolored Tangerine-Flake Streamline Baby*. New York: Farrar, Straus & Giroux, 1965.

————. *The Pump House Gang*. New York: Farrar, Straus & Giroux, 1968.

————. *Radical Chic and Mau-Mauing the Flak Catchers*. New York: Farrar, Straus & Giroux, 1970.

————. *The Right Stuff*. New York: Farrar, Straus & Giroux, 1979.

RESEARCH AND LITERARY THEORY

Anderson, Chris, editor. *Literary Nonfiction: Theory, Criticism, Pedagogy*. Carbondale and Edwardsville: Southern Illinois University Press, 1989.

Anderson, Chris. *Style as Argument: Contemporary American Nonfiction*. Carbondale and Edwardsville: Southern Illinois University Press, 1987.

Barthes, Roland. "Introduction to the Structural Analysis of Narratives," in *Image, Music, Text,* translated by Stephen Heath. New York: Hill and Wang, 1977.

————. *Mythologies*. New York: Hill and Wang, 1972.

Canary, Robert H., and Henry Kozicki. *The Writing of History: Literary Form and Historical Understanding*. Madison: University of Wisconsin Press, 1978.

Cheney, Theodore A. Rees. *Writing Creative Nonfiction*. Cincinnati: Writer's Digest Books, 1987.

Davis, Lennard. *Factual Fictions: The Origins of the English Novel*. New York: Columbia University Press, 1983.

Dennis, Everette E., and William L. Rivers. *Other Voices: The New Journalism in America*. San Francisco: Canfield Press, 1974.

Eason, David. "On Journalistic Authority: The Janet Cooke Scandal," in *Critical Studies in Mass Communication* 3: 429–47 (Dec. 1986). Also in *Media, Myths, and Narratives,* edited by James W. Carey. Newbury Park, Calif.: Sage Publications, 1988.

————. "New Journalism, Metaphor and Culture," in *Journal of Popular Culture* 15(4):142–48 (Spring 1982).

Fishkin, Shelley Fisher. *From Fact to Fiction; Journalism and Imaginative Writing in America*. Baltimore and London: The Johns Hopkins University Press, 1985; Oxford University Press paperback 1988.

Fletcher, Angus, ed. *The Literature of Fact*. New York: Columbia University Press, 1976.

Ford, Edwin H. *A Bibliography of Literary Journalism in America*. Minneapolis: Burgess Publishing Co., 1937.

Good, Howard. *Acquainted with the Night.* Metuchen: Scarecrow Press, 1986.

Hellman, John. *Fables of Fact: The New Journalism as New Fiction.* Urbana: University of Illinois Press, 1981.

Hersey, John. "The Legend on the License," in *The Yale Review,* 70(1):1–25 (Autumn 1980).

———. "The Novel of Contemporary History," in *Atlantic Monthly,* 184:80–84 (Nov. 1949). Lists the tasks required of the novel of contemporary history, many of which would double for literary journalism.

Heyne, Eric. "Toward a Theory of Literary Nonfiction," in *Modern Fiction Studies* 33:479–90 (Autumn 1987).

E. D. Hirsch, Jr. *Validity in Interpretation.* New Haven: Yale University Press, 1967.

Holland, Norman. "Prose and Minds: A Psychoanalytic Approach to Non-Fiction," in *The Art of Victorian Prose,* edited by George Levine and William Madden. New York; Oxford University Press, 1968.

Hollowell, John, *Fact and Fiction: The New Journalism and the Nonfiction Novel.* Chapel Hill: University of North Carolina Press, 1977.

Johnson, Michael L. *The New Journalism.* Lawrence: University of Kansas Press, 1971.

McCord, Phyllis Frus. *News and the Novel: A Theory and a History of the Relation between Journalism and Fiction.* New York University Ph.D. dissertation, 1985.

Mitchell, W. J. T. *On Narrative.* Chicago and London: University of Chicago Press, 1981.

Murphy, James E. "The New Journalism: A Critical Perspective," *Journalism Monographs* 34 (May 1974).

Olney, James. *Metaphors of Self: The Meaning of Autobiography.* Princeton: Princeton University Press, 1972.

Rockwell, Joan. *Fact in Fiction: The Use of Literature in the Systematic Study of Society.* London: Routledge & Kegan Paul, 1974.

Van Maanen, John. *Tales of the Field; On Writing Ethnography.* Chicago and London: University of Chicago Press, 1988.

Webb, Joseph M. "Historical Perspective on New Journalism," in *Journalism History* 1:38–60 (Summer 1974).

Weber, Ronald. "Journalism, Writing, and American Literature." Gannett Center for Media Studies Occasional Paper No. 5, Columbia University, April 1987.

———. *The Literature of Fact.* Athens: Ohio University Press, 1980.

———. *The Reporter as Artist: A Look at the New Journalism Controversy.* New York: Hastings House, 1974.

White, Hayden. "Introduction" to *The Content of the Form: Narrative Discourse and Historical Presentation.* Baltimore: Johns Hopkins University Press, 1987.

———. *Tropics of Discourse: Essays in Cultural Criticism.* Baltimore and London: The Johns Hopkins University Press, 1978.

Zavarzadeh, Mas'ud. *The Mythopoeic Reality: The Postwar American Nonfiction Novel.* Urbana: University of Illinois Press, 1976.

HISTORICAL SOURCES AND STUDIES

Anon., "Borderland of Literature," in *Spectator* 71:513 (Oct. 14, 1893).

Anon., "Confessions of a Literary Journalist," in *Bookman* 26:370–76 (Dec. 1907).

Anon., "Literature and Journalism," in the *New York Commercial Advertiser,* Jan. 1, 1897.

Anon., "Plea for Literary Journalism," in *Harper's Weekly,* 46:1558 (Oct. 25, 1902).

Anon., "The Point of View: The Newspaper and Fiction," in *Scribner's Magazine* 40:122–24 (July 1906).

Benson, Jackson J. *The True Adventures of John Steinbeck, Writer.* New York: The Viking Press, 1984.

Boylan, James. "Publicity for the Great Depression; Newspaper Default and Literary Reportage," in *Mass Media Between the Wars; Perceptions of Cultural Tension, 1918–1941.* Edited by Catherine Covert and John Stevens. Syracuse, N.Y.: Syracuse University Press, 1984.

Boynton, H.W. *Journalism and Literature and Other Essays.* Boston and New York: Houghton Mifflin & Co., 1904.

———. "The Literary Aspect of Journalism," in *Atlantic Monthly* 93:845–51 (June 1904).

Bruccoli, Matthew, editor. *Ernest Hemingway, Cub Reporter; Kansas City Star Stories.* Pittsburgh: University of Pittsburgh Press, 1970.

Chametzky, Jules. *From the Ghetto; The Fiction of Abraham Cahan.* Amherst: University of Massachusetts Press, 1977.

Coles, Robert. "James Agee's Search," in *Raritan* (Summer 1983), p. 74–100.

Connery, Thomas. "Hutchins Hapgood and the Search for a 'New Form of Literature,' " in *Journalism History* 13(1):2–9 (Spring 1986).

DeMott, Robert, editor. *Working Days: The Journals of the Grapes of Wrath.* New York: The Viking Press, 1989.

Eastman, Max. *Journalism Versus Art.* New York: Alfred A. Knopf, 1916.

Fenton, Charles A. *The Apprenticeship of Ernest Hemingway; The Early*

Years. New York: New American Library, 1961. Originally published 1954.

Gingrich, Arnold. *Nothing But People; The Early Days at Esquire: A Personal History 1928–1958*. New York: Crown Publishers, 1971.

Hapgood, Hutchins. "A New Form of Literature," in *Bookman* 21:424–27 (1905).

Hovey, Tamara. *John Reed: Witness to Revolution*. Los Angeles: George Sand Books, 1975.

Hughes, Helen M. *News and the Human Interest Story*. Chicago: University of Chicago Press, 1940.

Kaplan, Justin. *Lincoln Steffens, A Biography*. New York: Simon and Schuster, 1974.

Kramer, Dale. *Ross and The New Yorker*. Garden City, N.Y.: Doubleday & Co., 1951.

Kramer, Victoria A. *James Agee*. Boston: G. K. Hall, 1975.

Lee, Gerald Stanley. "Journalism as a Basis for Literature," in *Atlantic Monthly* 85:231–37 (Feb. 1900).

Lingeman, Richard. *Theodore Dreiser: At the Gates of the City, 1871–1907*. New York: G. P. Putnam's Sons, 1986.

Martin, Edward A. *H. L. Mencken and the Debunkers*. Athens: University of Georgia Press, 1984.

Moreau, Genevieve. *The Restless Journey of James Agee*. New York: Morrow, 1977.

Muggli, Mark Z. "The Poetics of Joan Didion's Journalism," in *American Literature* 59:402–21 (Oct. 1987).

Orhn, Karin. *Dorothea Lange and the Documentary Tradition*. Baton Rouge: Louisiana State University Press, 1980.

Preston, Richard McCann. *The Fabric of Fact: The Beginnings of American Literary Journalism*. Princeton University Ph.D. dissertation, 1983.

Rampersad, Arnold. *The Art and Imagination of W. E. B. Du Bois*. Cambridge: Harvard University Press, 1976.

Rischin, Moses, editor. *Grandma Never Lived in America; The New Journalism of Abraham Cahan*. Bloomington: Indiana University Press, 1985.

Sokolov, Raymond. *Wayward Reporter; The Life of A. J. Liebling*. New York: Harper & Row, 1980.

Steinbeck, Elaine, and Robert Wallsten, editors. *Steinbeck: A Life in Letters*. New York: The Viking Press, 1975.

Stephens, Robert O. *Hemingway's Nonfiction; The Public Voice*. Chapel Hill: University of North Carolina Press, 1968.

Stott, William. *Documentary Expression and Thirties America*. New York: Oxford University Press, 1973.

Weber, Ronald. *Hemingway's Art of Nonfiction.* New York: St. Martin's, 1990; London: Macmillan, 1990.

Wolfe, Tom. *The New Journalism.* With an anthology edited by Tom Wolfe and E. W. Johnson. New York: Harper & Row, 1973.

Ziff, Larzer. *The American 1890s: Life and Times of a Lost Generation.* New York: The Viking Press, 1966.

Contributors

TOM CONNERY is associate professor and chair of the Department of Journalism and Mass Communication at the College of St. Thomas in St. Paul, Minnesota. He is a former president of the American Journalism Historians Association. Connery has written on reporting and journalistic writing style in the late nineteenth and early twentieth centuries and is editing a reference guide to American literary journalism.

DAVID EASON is associate professor of communications at the University of Utah. He writes on issues in documentary representation, media history, and popular culture. His essays have appeared in the *Sage Annual Review of Communication Research, Critical Studies in Mass Communication, Communication Research, Journal of Popular Culture, Journal of Communication Inquiry,* and *Qualitative Sociology.* From 1987 through 1989 he was the editor of *Critical Studies in Mass Communication.*

SHELLEY FISHER FISHKIN is associate professor of American Studies at the University of Texas at Austin. She is the author of *From Fact to Fiction: Journalism and Imaginative Writing in America* (Johns Hopkins, 1985; Oxford University Press, 1988), which won the National Journalism Scholarship Society's Frank Luther Mott/Kappa Tau Alpha Research Book Award. She has published on journalism history and American literature in *The New York Times Book Review, American Journalism, American Literature,* and *The Colum-*

bia Journalism Review. Her current projects include editing (with Carla Peterson) an anthology of readings from the nineteenth-century African-American press and (with Elaine Hedges) a collection centered on Tillie Olsen's work called *Listening to 'Silences:' New Essays in Feminist Criticism.* She is working on books about gender and American nonfiction narrative, and about blurring the lines between journalism and fiction, history and the novel.

WILLIAM HOWARTH is professor of American Studies and English at Princeton University. He has written three books on Thoreau including *The Book of Concord* (1982), *Thoreau in the Mountains* (1982), and *The Literary Manuscripts of Henry D. Thoreau* (1974), as well as editing several compilations of Thoreau's work. He wrote *Traveling the Trans-Canada* (1987) and *Nature in American Life* (1972). He edited and wrote an introduction to *The John McPhee Reader* (1981). Howarth's scholarly articles, essays and journalism have appeared in *National Geographic, The Washington Post, The New York Times, Smithsonian, Sewanee Review,* and *New England Quarterly.*

HUGH KENNER is professor of English at Johns Hopkins University. His recent books include *Mazes* (1989), a collection of essays; *A Sinking Island* (1987), on modern English writers; *A Colder Eye* (1983), on modern Irish writers; and *The Mechanic Muse* (1987), on the effects of mechanization and urbanization on modern literature.

DARREL MANSELL is professor of English at Dartmouth College. He is the author of *The Novels of Jane Austen* and of many articles on Jane Austen, George Eliot, Matthew Arnold, William Shakespeare, Ernest Hemingway, and F. Scott Fitzgerald. He has also written on literary theory, the pianist Glenn Gould, the Don Juan myth in contemporary rock and roll, and an autobiographical piece published in *Italian Quarterly.*

MARY MCCARTHY was the author of, among other things, *The Groves of Academe* (1952), *Memories of a Catholic Girlhood* (1957), *The Group* (1963), *Vietnam* (1967), *Hanoi* (1968), *Cannibals and Missionaries* (1979), and *How I Grew* (1987). Her articles for this volume came from a collection of essays called *On the Contrary* (1961). She taught literature at several institutions, including Bard College and Sarah Lawrence College.

JOHN PAULY's articles on the history and sociology of the mass media have appeared in *Critical Studies in Mass Communication, American Quarterly, American Journalism, Communication Research,* and other journals and books. He is associate professor of communication at the University of Tulsa, and the editor of *American Journalism.*

NORMAN SIMS is associate professor and chair of the Journalism Department at the University of Massachusetts. He edited and wrote an introduction to *The Literary Journalists* (Ballantine, 1984) and his articles have appeared in *Journalism History, The Quill, Critical Studies in Mass Communication, Gannett Center Journal,* and *The Dictionary of Literary Biography.*

KATHY SMITH is assistant director of the New Hampshire Humanities Council. She has recently returned to her home state after several years and a Ph.D. in English from the University of Massachusetts. During her academic exile, she taught composition, American fiction and journalism at North Adams State College, the University of Massachusetts and St. Lawrence University in New York. She is currently writing grants, poetry, and thinking about revising her dissertation entitled, "Writing the Borderline: Journalism's Literary Contract."

RON WEBER is professor of American Studies at the University of Notre Dame. His books include *The Reporter as Artist* (1974) and *The Literature of Fact* (1980), which won a Choice award for best academic book of the year. *Hemingway's Art of Nonfiction* was published by St. Martin's Press in 1990 in New York, and by Macmillan in London. His articles on nonfiction have appeared in *Virginia Quarterly, Antioch Review, Sewanee Review,* and *The Journal of Popular Culture.* In 1985–1986 he was a research fellow at the Gannett Center for Media Studies at Columbia University.